Understanding Networking Technology

Concepts, Terms, and Trends

Second Edition

For a complete listing of the *Artech House Telecommunications Library,*
turn to the back of this book.

Understanding Networking Technology

Concepts, Terms, and Trends

Second Edition

Mark Norris

Artech House
Boston • London

Library of Congress Cataloging-in-Publication Data
Norris, Mark.
Understanding networking technology: concepts, terms, and trends / Mark Norris—2nd ed.
p. cm.—(The Artech House telecommunications library)
Includes bibliographical references.
ISBN 0-89006-998-0 (alk. paper)
1. Telecommunication—Dictionaries. 2. Computer networks—Dictionaries.
I. Title. II. Series.
TK5102.N67 1999
004.6'03—dc21 99-26560
CIP

British Library Cataloguing in Publication Data
Norris, Mark
Understanding networking technology: concepts, terms, and trends—2nd ed.
1. Telecommunication—Dictionaries
I. Title
621.3'82

ISBN 0-89006-998-0

Cover and text design by Darrell Judd

International Standard Book Number: 0-89006-998-0
Library of Congress Catalog Card Number: 99-26560

10 9 8 7 6 5 4 3 2 1

Contents

ABCDEFGHIJKLMNOPQRSTUVWXYZABCDEFGHIJKLMNOPQRSTUVWXYZABCDEFGHIJKLMNO

Acknowledgments

Many people have contributed to this tome through volunteering a piece of information, a better definition, or a logical link. A few have done more than their fair share and need to be named! I would like to thank Dr. Allan Hudson and Steve West, who found, reviewed, and checked many of the trends presented at the end of the book, and Chris Stannard, whose attention to detail has done much to ensure accuracy in the glossary.

I am also most grateful for all the help and encouragement from Julie Lancashire, Kate Hawes, and Susanna Taggart from Artech Books in London, and Tina Kolb in Boston. Finally, I am indebted to, and most impressed by, the anonymous reviewers used by Artech, who added significantly to the content and cogency of this work.

Introduction

ABCDEFGHIJKLMNOPQRSTUVWXYZABCDEFGHIJKLMNOPQRSTUVWXYZABCDEFGHIJKLMNO

Tyranny begins with a corruption of the language.
—W. H. Auden

With technology pervading more and more areas of modern life, its associated vocabulary has grown ever more difficult for the nonspecialist to grasp. In particular, the convergence of computing and telecommunications has led to a broad and often confusing set of terms that are assumed by many suppliers, consultants, and technocrats to be widely known.

We list here many of the abbreviations and concepts in common use these days. The main aim of this listing is to clarify some of the more complex ideas and terms (often liberally bandied around) by giving some idea of their context. The explanations below are intended to help specialists, generalists, hobbyists, or observers who deal with a wide range of technologies to interpret what they see and hear.

There should be just enough information here to enable a reader to ask sensible questions and have a reasonable understanding. The coverage is far from complete—that was not the intent of this text. Rather, the aim is to provide some words that will firmly fix some of the more central ideas in the reader's mind.

This text started life as an extended crib sheet. In its early days it was not intended to do any more than aid in the communication between a collection of people from disparate backgrounds. The first draft was only a few pages long. But success followed, and the popularity of the small text soon led to the production of a more complete one. Little by little the text expanded as more and more people saw a path to instant erudition.

The current version is an attempt to balance usefulness with completeness. The brevity of presentation is intended to make casual browsing both useful and rewarding—the reader should, with minimal effort, be able to store up a good set of concepts for later reference. This book contains all of the main terms and concepts that the reader could reasonably be expected to be aware of. Needless to say, there will be terms that do not appear in this version. The sheer rate of change and breadth of this area of technology mean that we will probably never catch up with all the new vocabulary. The acid test I have tried to pass is whether there is enough here to get readers where they want to go.

I have included a section that builds on established, historic information to provide a feel for the overall trends in technology and speculation on what the future might hold. This final part of the book is intended to give some sort of context. It covers a wide range of technical aspects, from the inexorable growth in processing power to the ways in which technology impacts everyday life. The picture that emerges has been built from many different inputs, so it should be taken as no more than an informed guess. Even so, it should be clear from this section that some things are reasonably predictable, while others are definitely still very transient. It is well worth knowing something of the big picture before delving into the details of a particular technology.

A set of established principles and benchmarks would have fit very well into this book. It would have been very nice to be able to state categorically how much a line of code costs, what the communications requirements for a teleworker are, or where the intelligence in a distributed system should be placed. Unfortunately, a section containing this kind of material defies sensible compilation. The closest thing that can be offered is the section on general trends for computers and communications.

With the ever-increasing need for everyone to understand a little of everyone else's business, many have found this short text (for all of its imperfections, omissions, and subjectivity) useful in bringing a complex world of technology into useful focus. After all, no one can really comprehend all that goes on in computing science, software, and communications. The ability to abstract from precise detail is vital; the spirit of this guide can be summarized in the words of H. H. Munro (Saki): "A little inaccuracy sometimes saves tons of explanation."

One final point to make is that this book is very deliberately committed to paper rather than being online. Having spent several years as part of a "virtual" team, I know how useful information on a computer screen can be. I also know its limitations. The convenience and portability of paper provides an ideal medium for this sort of work. The prerequisites of special training and equipment that go with any online information are the very things I have tried to avoid.

Note

The terms that are italicized in the definitions are themselves defined as separate entries.

ABCDEFGHIJKLMNOPQRSTUVWXYZABCDEFGHIJKLMNOPQRSTUVWXYZABCDEFGHIJKLMNO

AAL

Asynchronous transfer mode adaptation layer. A series of protocols that enable *ATM* to be made compatible with virtually all of the commonly used standards for voice, data, image, and video. It is the essential "glue" that connects a high-speed technology with the information it carries.

There are four adaptation layers, ranging from AAL1 (specialized for services such as voice) to AAL5 (intended as a simple and efficient carrier for data services). AALs 3 and 4 were merged early on in the application of ATM.

AAL1

The ATM adaptation layer designed to support connection-oriented service. It has a 1-byte overhead containing a sequence number to detect missing cells.

AAL2

The ATM adaptation layer that, like AAL1, is designed to support connection-oriented service. The main difference is that AAL2 caters for variable, rather than continuous bit rates.

AAL3

The ATM adaptation layer designed for connection-oriented data. Originally intended for services like frame relay and signaling, it was subsequently combined with the similar AAL4 to yield the AAL3/4 now in use.

AAL5

Known as efficient AAL, AAL5 has become the accepted layer for frame relay and signaling. It is also under study as the AAL for protocols such as IP and LAN networks such as Novell's IPX.

Abends

An "abnormal end" is often referred to as an abend, the term most widely used for what happens when a transaction has not completed successfully. Often there will be additional error codes displayed on an abend screen which more clearly pinpoints where in the transaction process the transaction actually failed.

ABI

Application binary interface. An interface by which *application programs* gain access to an *operating system* and other services. It enables the producers of *application programs* to write their code independently of a specific machine.

AB signaling

A type of signaling used with *T1* transmission circuits. The least significant bits from the 6th and 12th subchannels of the 24-channel T1 frame are robbed to carry data and control information. The former is the A bit, the latter the B bit.

This technique can be extended to ABCD signaling by robbing the 18th and 24th subchannels, although this is rarely done in practice.

Absolute address

A reference to a storage location that is in terms of a fixed displacement from memory location zero, rather than any relative location.

Abstraction

A representation of some object that contains less information than the object itself. For example, a data abstraction provides information about some reference in the outside world without indicating how the data are represented in the computer.

Access control

A general term that relates to the hardware, software, and administrative tasks that manage the interface between a user and a network. In general, access control covers user identification, authorization, and privilege.

Access control method

Usually relates to the selection of the appropriate form of communications in a *LAN*. By regulating a workstation's physical access to the transmission medium, it directs traffic around the network and determines the order in which nodes gain access so that each user obtains an efficient service. The most commonly found access methods include *token ring*, *FDDI*, and *CSMA/CD*.

Access control method is also used to describe the means of providing input/output in a computer operating system. It usually implies a data or file structure that is used to communicate with an external entity, such as a printer or other peripheral.

Access line

Connection from a private user to a public network (e.g., the public switched telephone network or a data network). Also known as a local line or local loop.

Access time

The average time interval between a storage peripheral (usually a disk drive or memory) receiving a request to read or write a certain location and returning the value read or completing the write. In order for a disk drive to start to read or write a given location, a read/write head must be positioned radially over the right track and rotationally over the start of the right sector. The radial motion (sometimes called seeking) is what causes the intermittent noise heard when a hard disk is being accessed.

Accumaster

A service offered by *AT&T* for managing a user's network. It usually relates to a customer with a large, complex network that spans a wide area.

ACD

Automatic call distributor. A system designed to distribute incoming telephone calls among a number of operators. Usually used in *call centers* to allow a large number of telephone queries to be answered by a relatively small number of operators. ACDs are frequently used by organizations offering services such as telephone support desks, inquiry bureaus, information desks, and reservation assistance.

ACF

Advanced communication function. General term for a series of software products from IBM that allows networks of computers and applications to be built.

ACID

An acronym for the key properties that should be sought in the design of distributed systems. The ACID properties are atomicity, consistency, isolation, and durability.

ACM

Association for Computing Machinery. Founded shortly after the Second World War, this is the largest and oldest of the industrial scientific and educational computer societies. The ACM has some 90,000 members, provides a broad forum for monitoring technical developments, and has a wide variety of publications, special interest groups, chapters, conferences, and awards.

Acrobat

A widely used and general purpose application from Adobe, composed of three different products: Acrobat Reader, Acrobat Exchange, and Acrobat Distiller. They use Adobe's portable document format (PDF), a PostScript-based file format than can describe documents in a completely device- and resolution-independent manner.

ACSE

Association control service element. An *ISO* standard that deals with the setup and clear-down of a dialog between applications on two remote machines. The session between the two applications is, in ISO parlance, known as an association, hence the name.

Active-X

Microsoft technology for embedding information objects and application components within one another. For example, an Active-X button can be embedded in an HTML page that is displayed in a browser window.

Ada

A programming language, developed under the auspices of the U.S. Department of Defense, specifically oriented to support advanced software development practices such as abstraction and reusability. The original language was standardized as Ada 83 and the most recent update was Ada 95. It is a large, complex, block-structured language aimed primarily at large-scale and embedded applications.

Address

A common term used both in computers and data communications that designates the destination or origination of data or terminal equipment in the transmission of data. Types of address include hardware addresses (e.g., 0321.6B11.5643, for an Ethernet card), logical address (e.g., 131.146.6.11, a TCP/IP address for a work- station), and a personal address (e.g., M.Norris@ axion.bt.co.uk, to reach an individual).

The term is also used to denote a location that can be referred to in a piece of software.

Address mask

A networking term that is also referred to as a subnet mask. It is used to identify which bits in an *IP* address correspond to the network address and which bits refer to the *subnet* portion of the address.

Address resolution

A networking term that refers to the conversion of an Internet address into the corresponding physical address (e.g., a corresponding *Ethernet* address).

Address space

The range of addresses that a processor can access. This usually depends on the width of the processor's address bus and address registers. Address space may refer to either physical addresses or virtual addresses.

ADI

Awareness, Decisions, and Implementation. Or in full, Awareness Creation, Decision Making, and Support and Decision Implementation. A concept used within the network management community for describing the essential sequence of activities in managing a network.

ADM

Add Drop Multiplexer, a device for extracting specific data streams from a multiplex stream. ADMs are often used with SDH systems to extract or include a tributary.

ADMD

Administrative Management Domain. Part of the X.400 messaging standard, ADMD represents a public messaging service such as CompuServe.

Administrative domain

Used to denote a logical network group managed as a single resource. Usually comprises a collection of hosts, routers, servers, and the interconnecting network(s) managed by a single administrative authority.

ADPCM

Adaptive differential pulse-code modulation. A standard coding technique that allows voice to be transmitted over a 32-Kbps channel, instead of the standard 64-Kbps channel required by *PCM* coding. Sampling is done at the conventional rate of 8 kHz, but each sample is encoded as 4 bits that describe the difference from the adjacent sample (rather than 8 bits that describe the sample itself).

ADSI

Analog display services interface. A signaling protocol developed by *Bellcore* that allows a user to manipulate and control advanced services. ADSI can be seen as the next step up from *CLASS.*

ADSL

Asymmetric digital subscriber line. A digital telecommunications protocol that allows data of speeds up to 6 Mbps to be transmitted over local telephone lines. At a very basic level, ADSL works by dividing the capacity available over the telephone line into a number of frequency bands and using each as a data path. As the name suggests, the transmission rate is not the same in each direction. There is usually a high-speed downstream path (typically 1.5 to 6.4 Mbps) with a much slower (16 to 64 Kbps) upstream return.

This makes ADSL an important technology in the provision of multimedia services over existing phone lines, since high-bandwidth applications (e.g., video) can be delivered under consumer control (which requires minimal bandwidth for signaling). That ADSL uses frequencies from just above the voice band (about 4 kHz) up to around 900 kHz means that the reach over ordinary copper wire is limited. A distance of 3 km is typical for data rates of 2 Mbps.

ADT

Asynchronous Data Transfer, a method of data transfer used by SCSI systems. This is the type of transfer rate originally introduced with SCSI-1 that commonly provides transfer rates of 2 MBps.

AFIPS American Federation of Information Processing Societies.

AFNOR Association Francaise de Normalisation, the French national standardization body.

AFP AppleTalk Filing Protocol, that lets workstations access files from remote file servers. The protocol corresponds to layer 6 of the Open Systems Interconnection (OSI) reference model.

AGCH The Access Grant Channel, part of the GSM standard for mobile communications.

Agent A piece of software that carries out a particular set of predefined tasks. For instance, a mail agent might be installed on a *PC* to monitor and filter incoming messages.

Aggregate The total bandwidth of a multiplexed bit stream channel, expressed as bits per second.

AI Artificial intelligence. Applications that would appear to show intelligence if they were carried out by a human being. This was one of the hottest topics in computing during the 1980s and promised to lead the way to a new raft of computer applications. The difficulties of defining and coding intelligence have proved more challenging than was initially predicted.

AIN Advanced intelligent network. A version of the intelligent network, *IN*, based on specifications and products from *Bellcore*.

AIX The name given to the Unix operating system ported onto IBM computers, such as the RS/6000 series.

A-law An *ITU-T* standard for the digitization of an analog signal. It is a nonuniform quantizing logarithmic compression method commonly used across Europe as a means of encoding speech on a public telephone network (an alternative to *mu-law* encoding).

ALGOL 60

Algorithmic language. One of the earliest of the modern (third-generation) programming languages. Introduced in 1960, it was a portable language for scientific computations. ALGOL 60 was small and elegant. It was block-structured, nested, recursive, and freeform. It was also the first language to be described in *BNF.*

Algorithm

A group of defined rules or processes for solving a problem. This might be a mathematical procedure enabling a problem to be solved in a definitive number of steps or a precise set of instructions for carrying out some computation (e.g., the algorithm for calculating an employee's take-home pay).

Aloha

An experimental packet-switched network developed at the University of Hawaii in the mid-1970s. Although radio-based, Aloha set many of the principles for subsequent commercial *packet networks.*

Alpha test

An early test of an entire system. Usually, a controlled release to a small number of cooperative individuals or sites.

Amdahl's Law

A means of calculating the speed benefits of using processors in parallel. It states that if F is the fraction of a program that is sequential and $1 - F$ is the fraction that can be executed in parallel, then the speed benefit of using P processors is $1/\{F + [(1 - F)/P]\}$. This law shows quite clearly where the limits lie when adding more computing power to a problem.

AMI

Alternate mark inversion. A digital signaling method (also known as bipolar) designed to ensure that there are no long sequences of inactivity on a transmission line (and hence risk of losing synchronization). AMI is used for both *T1* and *E1* lines.

AMPS

Advanced Mobile Phone Service. This was the name given to a U.S.-based mobile phone system provided by *AT&T.*

AN

Army/Navy. U.S. military electronics equipment, which is specified in the Joint Electronics Type Designation System, is identified with the AN prefix.

The full AN series is extensive and covers many types of radio and field telephony equipment.

Analog

In electronics, used to describe a continuously variable signal, as opposed to a discrete or digital one (e.g., 0s or 1s). In general, analog circuits are harder to design and analyze than digital ones because account must be taken of effects such as the gain, linearity, and power rating of components; the resistance, capacitance, and inductance of wires and connectors; interference between signals; power supply stability; and electromagnetic interference.

Digital circuit design, especially for high-speed switching, must also take these factors into account, but they are usually less critical because most digital components will function correctly within a wide range of conditions, whereas even small variations tend to corrupt the outputs of an analog circuit.

Analog computer

A machine or electronic circuit designed to work on numerical data represented by some physical quantity (e.g., rotation or displacement) or electrical quantity (e.g., voltage or charge) that varies continuously, in contrast with digital signals (which are either 0 or 1). Analog computers are said to operate in real time and are used for research in designs where many different shapes and speeds can be tried out quickly. For instance, an analog computer model of a car suspension allows the designer to see the effects of changing size, stiffness, and damping.

Anchor

A named location in a hypertext page or document that gives the destination for a hypertext link. When the user clicks on a piece of link text or image, the system loads a new page and displays the section identified in the anchor. In HTML, an anchor is denoted thus:

<A NAME="where you want to go"> </A>

ANDF

Architecture-neutral distribution format. A specified format for software from the OSF that enables the distribution of a single version of an *application* to computers with different hardware and software architectures.

Anisochronous

A signal that is not related to any clock. Transmission can occur at any time.

Anonymous FTP

An interactive service provided by many *Internet* hosts allowing users to transfer documents, files, programs, and other archived data. Users log in using the special user name "ftp" or "anonymous" with their *e-mail* address as password. They then have access to a special directory hierarchy containing all of the publicly accessible files. This is generally a separate area from the private files used by local users of that machine.

ANSA

Advanced Networked Systems Architecture. A research group established in Cambridge, U.K., in 1984 that has had a major influence on the design of distributed processing systems. The ANSA reference manual is probably the closest thing to an engineer's handbook in this area. The reference software produced by the group has provided the basis for a number of commercial distributed systems.

ANSI

American National Standards Institute. A U.S.-based organization, affiliated with the *ISO*, which develops standards and defines interfaces for telecommunications.

The standard version of the C programming language, ANSI-C, is probably the best known output. ANSI standards for *EDI* are also widely known (X.12 for general EDI and X.9.9 for message authentication for banking applications).

Some of the better known and widely used ANSI specifications are:

ANSI X3.124	Graphics standard, equates to ISO 7942
ANSI X3.135	SQL data access standard, equates to ISO 9079-1989
ANSI X3.138	Repository standard, IRDS
ANSI X3.143	SGML standard for documents
ANSI X3.144	Graphics standard that equates to ISO 9542

ANSI X3.64	Terminal emulation standard
ANSI X3T2	Syntax notation, same as ASN.1 ISO 8824-5
ANSI X3T9	Fiber Distributed Data Interface (FDDI)-physical layer protocol.

Antenna

Device used to transmit and/or receive radio waves. The physical design of the antenna determines the frequency range of transmission/ reception.

APCN

Asia Pacific Cable Network. One of a number of regional undersea telecommunications networks. APCN connects Japan, Korea, Singapore, and others with an 11,500-km trunk and branch cable.

APDU

Application protocol data unit. A packet of data exchanged between two *application programs* across a network. This is the highest level view of communication in the *ISO* seven-layer model, and a single packet exchanged at this level may actually be transmitted as several packets at a lower layer, as well as have extra information (headers) added for routing and other network-specific requirements.

API

Application program interface. The interface by which an *application program* accesses operating systems and other services. An API is usually defined at the source code level in the form of defined calling conventions, available resources, and data. It provides a level of abstraction between the application and the underlying machine. APIs are defined to ensure the portability of the code and to encourage third-party software providers to write applications for systems.

APIs are usually built as software that makes the computer's facilities accessible to the *application program.* All operating systems and network operating systems have APIs. In a networking environment, it is essential that various machines' APIs are compatible; otherwise, programs would be exclusive to the machines in which they reside.

APOP

An encryption scheme to pass passwords over the net while logging into a POP3 server. With APOP, only the password is encrypted, not the mail itself.

APPC

Advanced Program-to-Program Communication. An *API* developed by IBM. Its original function was in mainframe environments, enabling different programs on different machines to communicate. As the name suggests, the two programs talk to each other as equals, using APPC as an interface to ensure that different machines on the network talk to each other.

APPC/PC

A version of the *APPC* developed by IBM to run a *PC*-based *token ring* network.

Appleshare

The software for the *Macintosh* environment that allows the sharing of files between Mac users.

Applet

A small *application program* that can be accessed over a network (typically the *Internet*). It is self-contained in that it carries its own presentation and processing code and can run on whatever type of machine imports it. Although fairly new, applets are being used as "plug-in" units that form part of a larger application.

The concept of the applet is tied to that of *Java* (a compact and portable language) and *Hot Java* (a *browser* similar to others used on the *World Wide Web*).

AppleTalk

A set of ISO-compliant protocols that are independent of the underlying media and able to run on *Ethernet*, *token ring*, and LocalTalk (Apple Computer's proprietary cabling system for connecting *PCs*, *Macintoshes*, and peripherals using the *CSMA/CA* access method).

Application

The user task performed by a computer, such as making a hotel reservation, processing a company's accounts, or analyzing market research data.

Application generator

High-level language that allows rapid generation of executable code, sometimes referred to as *4GL*. Focus is a typical example.

Application layer

Layer 7 of the OSI reference model; implemented by various network applications, including electronic mail, file transfer, and terminal emulation.

Application program

A complete, self-contained program that performs a specific function directly for the user, in contrast with system software such as the operating system kernel, server processes, and libraries, which exist to support application programs. It is usually referred to simply as an "application."

Editors, spreadsheets, and text formatters are common examples of applications. Network applications include the software clients used to give users access to services such as *FTP*, *e-mail*, and *Telnet*. The term is used fairly loosely; for instance, some might say that a client and server together formed a distributed application; others might argue that editors and compilers are not applications but tools for building applications.

One distinction between an application program and the operating system is that applications always run in user mode only, and this confers no special privileges. In contrast, operating systems and their related utilities may run in supervisor or privileged mode and can effect changes to the system and anyone on it.

Application software

The software used to carry out a specific application task (e.g., a document editor, compiler).

APPN

Advanced Peer-to-Peer Networking. A facility with the IBM/SNA that provides distributed processing based on Type 2.1 network nodes and LU 6.2

APS

Asynchronous protocol specification. An *ITU* standard for remote access to *X.400* mail systems. It is used to send and receive mail over dialup access networks.

Archie

A tool that allows users to find information stored on a remote machine. Archie is usually associated with the *Internet* and works by locating files on *anonymous FTP* sites (of which there are many). To use Archie, users have to know the precise name of the file they are looking for.

Architecture

In computer and communication systems, denotes the logical structure or organization of the system and defines its functions, interfaces, data, and procedures. In practice, architecture is not one thing but a set of views used to control or understand complex systems. A loose (but useful) definition is that architecture is a set of components and some rules for assembling them.

Architecture style

A set of components, topological layout, set of interaction mechanisms, environment, and possibly technology (e.g., CORBA).

ARP

Address resolution protocol. A networking protocol that provides a method for dynamically binding a high-level *IP* address to a low-level physical hardware address. This means, for instance, finding a host's *Ethernet* address from its *Internet* address.

The sender broadcasts an ARP packet containing the Internet address of another host and waits for it (or some other host) to send back its Ethernet address. Each host maintains a *cache* of address translations to reduce delay and loading.

ARP allows the Internet address to be independent of the Ethernet address, but it only works if all hosts support it. ARP is defined in *RFC* 826. ARP also allows mapping from an Internet address to a *token ring MAC* address.

ARPANET

Advanced Research Projects Agency Network. The precursor to the *Internet.* It was initiated in the late 1960s as an experiment in resilient networking. Many of the current Internet standards derive from this early work.

ARQ

Automatic repeat request. A general approach to error control in transmission systems. If an error is detected by a receiver, it requests the transmitter to send the message again. There are various protocols that can be used with this approach to optimize overall speed or efficiency.

ASCII

American Standard Code for Information Interchange. The predominant character set used for coding in present-day computers. The modern version uses 7 bits for each character, whereas most earlier codes (including an early version of ASCII) used fewer. This extension allowed the inclusion of lowercase letters—a significant step forward—but it did not provide for accents and other special forms not used in English.

A commonly used character set is necessary because computers are much less flexible about spelling than humans. Thus, the way characters are defined and used has to be very precise.

ASE

Application service element. Software in the presentation layer of the *OSI* seven-layer model that provides an abstracted interface layer to *APDUs*. Because applications and networks vary, ASEs are split into common services and specific services. Examples of services provided by the common ASE (*CASE*) include remote operations (*ROSE*) and database *CCR*. The specific ASE (*SASE*) provides more specialized services.

ASN.1

Abstract Syntax Notation One. An ISO/CCITT standard language for the description of data. ASN.1 is defined in the *ITU* standard X.208 (which is equivalent to ISO standard 8824).

ASN.1, along with some standard basic encoding rules, facilitates the exchange of structured data between *application programs* over networks by describing data structures in a way that is independent of machine architecture and implementation language.

OSI application layer protocols such as *X.400 MHS* electronic mail, X.500 directory services, and *SNMP*, use ASN.1 to describe the *PDUs* that they exchange.

Assembler

A program, usually provided by the computer manufacturer, to translate a program written in *assembly language* into *machine code*. In general, each assembly language instruction is changed into one or two corresponding machine-code instructions.

Assembly

The process of converting a program written in *assembly language* into *machine code*.

Assembly language

A low-level programming language, generally using symbolic addresses and involving the manipulation of the processor registers, which is translated into *machine code* by an assembler. Growing system complexity precludes the practical use of this very-low-level language for many applications. It is not uncommon, though, for a part of a program to be written in *assembler*—especially when speed is important.

Assessor

A person who is qualified and is authorized to perform all or any portion of a quality system assessment. The term "auditor" can also be used.

As we may think

Visionary article published in 1945 by Vannevar Bush, science advisor think to President Eisenhower. It predicted many of the elements of the Information Age, including hypertext and the Internet.

Asynchronous

Contrasts with *synchronous* in that there is no correlation between data items and system time. Each character is transmitted as a separate entity.

Asynchronous transmission

A data transmission in which receiver and transmission transmitter clocks are not synchronized. Each character (word/data block) is preceded by a start bit and terminated by one or more stop bits, which are used at the receiver for synchronization.

Async-Sync PPP Conversion

Method by which PPP data sent between a computer's COM port and the ISDN are converted by the terminal adapter to/from asynchronous to synchronous traffic.

AT&T

American Telephone and Telegraph. Prior to its breakup (see *MFJ*), the largest telecommunications company in the world. Still a leading innovator in many areas of communications.

AT commands

The commands used to control a standard V.22 full duplex dialup modem from a personal computer (e.g., turn echo on/off, go off-hook, use pulse dialing). Also known as Hayes commands, after one of the leading manufacturers of intelligent modems.

ATM

Asynchronous transfer mode. A means of transporting high-speed data over networks based on uniform cell transmission. Equally applicable to local and wide areas and heralded as the uniform fabric for carrying voice, video, and data traffic. It is possible that ATM networks could provide the multiservice networks that would be required to underpin widely distributed multimedia applications.

ATM uses short, standard format, 53-byte cells to divide data into efficient, manageable packets for ultrafast switching through a high-performance communications network. The 53-byte cells contain 5-byte destination address headers and 48 data bytes. Various *AALs* are defined for ATM to allow different types of data to be carried effectively.

ATM is the first packet-switched technology designed from the ground up to support integrated voice, video, and data communication applications. It was originally conceived as the broadband extension to *ISDN* and is, therefore, suitable for use over *WANs*.

ATM currently accommodates transmission speeds of from 64 Kbps to 622 Mbps, but may well support gigabit speeds in the future. As it is, it promises to provide the uniform data transport technology of the twenty-first century.

ATM

Automatic teller machines. "Through-the-wall" cash dispensers. ATM is a good example of an overloaded acronym, freely used and sometimes misunderstood!

ATM Forum

A technical group, composed mainly of network equipment vendors, that has set itself the target of developing a set of pragmatic standards for ATM equipment.

ATN

The global Aeronautical Telecommunications Network, a satellite airline communications consortium comprising Telenor, SingTel, and BT, with a worldwide network that allows air traffic controllers to track long-haul commercial aircraft. The system is intended to alleviate the problem of aircraft being lost on radar as they cross large oceans.

ATRAC

The Sony Corporation's Adaptive Transform Acoustic Coding. It is used on the MiniDisc system and is similar to the PASC system developed by Philips. It reduces storage to about 20% of that required on a conventional CD, by thresholds and masking detection and sub-band coding.

Attachment

A file that is sent and received along with an email message. The file may be binary or text and is opened using an appropriate application.

Audit trail

A continuous record of a network's activity. This is a useful network management tool, as it shows how resources are being used and where the problems lie.

Authoring

The process of developing a multimedia application. The composite tasks of authoring include overall design, scripting, linking, and presentation.

Automata

Systems that can operate with little or no human intervention. It is easiest to automate simple mechanical processes, and hardest to automate those tasks needing common sense, creative ability, judgment, or initiative in unprecedented situations.

Automatic repeat request

A technique for minimizing the effects of errors in transmission systems. Usually referred to as ARQ, this technique allows several blocks of data to be sent before acknowledgment is required.

Availability

An important measure of the performance of a network or system. Usually expressed as a percentage, with 100% being the ideal. Availability is determined by the *MTBF* and the *MTTR*. The formal measure is MTBF/(MTBF + MTTR).

Avatar

A pseudopersonality adopted by a someone using a network to interact with others. Usually takes the form of a picture (chosen from a gallery) that can be used to represent an individual when interacting with graphical applications such as the *World Wide Web*.

AVI

A file format for interleaved audio and video. Introduced in 1990 by Microsoft and now an industry standard. The interleave ratio can be varied—for instance a ratio of 7 indicates that seven video frames separate each audio section.

ABCDEFGHIJKLMNOPQRSTUVWXYZABCDEFGHIJKLMNOPQRSTUVWXYZABCDEFGHIJKLMNO

Backbone
The top level in a hierarchical network. Stub, transit, and other networks that connect to the same backbone are guaranteed to be interconnected.

Backbone site
An *Internet* term used to describe any of the key *Usenet*, *e-mail*, and/or Internet sites, typically one that processes a large amount of traffic.

Backoff
A host that has experienced a collision on a network waits a (random) amount of time before attempting to retransmit. The idea of backing off is most popularly used in the *CSMA/CDU* access mechanism used by Ethernet.

BACS
Bankers Automatic Clearing System, the U.K. system used for clearing checks.

Baldridge
A prestigious award that is given on the basis of an organization's commitment to quality improvement. The criteria for the Baldridge award ranges from process control to company results to environmental policy. The award is named after Malcolm Baldridge, a former U.S. secretary of commerce. The European Quality Award (EQA) is the Euro-equivalent.

Balun
An abbreviation of balanced-unbalanced, it is an impedance-matching device that connects a bal-

anced line (such as a twisted-pair line) with an unbalanced line (such as a coaxial cable).

Bandwidth

The amount of information that can be sent over a given transmission channel. Usually expressed in terms of the range of frequencies or the number of bits per second that can be carried. Telephone lines usually have a bandwidth of around 4 kHz. The digital data that can be carried on the same wire vary according to the transmission technology used. (Two channels at 64 Kbps is typical, but much higher rates can be achieved with *ADSL*.)

Bandwidth is one of a number of ways of characterizing a network. Other important parameters that often need to be considered are *latency* and reach.

Baseband

A transmission medium through which signals are sent in their original form without any frequency shifting. In general, only one communication channel is available at any given time. *Ethernet* is an example of a digital baseband network. A telephone is a baseband analog instrument.

BASIC

Beginner's all-purpose symbolic instruction code. A simple language created in the early 1960s, which was designed for quick and easy programming by students and beginners. BASIC exists in many dialects and is popular on microcomputers with sound and graphics support. Most micro versions are interactive and interpreted (the source code is translated into machine code one line at a time), but the original BASIC was compiled.

Bastion host

A machine placed on the perimeter of a data network to provide publicly available services. Although secured against attack, it is assumed to be compromised because it is exposed to the Internet.

Batch

A term describing a system that takes a set (or a batch) of commands or jobs, executes them all in sequence, and returns the results without human intervention.

This contrasts with an interactive system in which the user's commands and the computer's responses are interleaved. A batch system typically takes its commands from a disk file (or a set of punched cards in the old days) and returns the results to a file (or prints them).

Often there is a queue of jobs that the system processes as resources become available.

Batch processing

In data processing or data communications, an operation in which related items are grouped together and transmitted for common processing.

Bathtub curve

Common term that describes the expected failure rate of electronics and other hardware devices with time: initially high, dropping to near zero for most of the system's lifetime, and then rising again as it wears out. The exact shape of the curve is determined by the *MTBF* for the device.

A similar cycle of initial problems followed by a period of stability followed by creeping problems has been observed in software and computer systems. This is sometimes referred to as the software death cycle (in contrast to the more familiar software *life cycle*).

Baud

The maximum information-carrying capacity of a communication channel in symbols (state transitions or level transitions) per second. This coincides with bits per second only for two-level modulation with no framing or stop bits. A baud may carry between 1 and 16 bits depending on the coding that is used.

The term "baud" was originally used as a unit of telegraph signaling speed—one Morse code dot per second. The name derived from a M. Baudot (1845–1903), the French engineer who constructed the first successful teleprinter and gave his name to an early coding standard.

BBS

Bulletin board system. A computer-based meeting and announcement system that allows people to both post and view information. It is often organized into user groups or centered around a particular subject. There are many *Internet*-based BBSs.

B channel

In the ISDN, the B channel provides the basic rate access of two 64-Kbps data paths in Europe and two 56-Kbps data paths in North America. This is fully described in *ITU-T* recommendation I.420.

The name originates from having the designation A for analog channels and B for digital channels.

Beacon

A feature in a token ring frame signaling system that indicates that the ring is inoperative because of a serious hard error—a defective cable or faulty nodes for instance.

Bearer

Generally used to refer to a transmission channel that is used to carry some data. For instance, in *ISDN*, there is a 64-Kbps bearer channel available. It can be used to carry a variety of data at speeds up to and including 64 Kbps.

BECN

Backward Explicit Congestion Notification standard, used in frame relay and ATM. Not effective when congestion is of short duration, or if some terminals are unable to react.

Bellcore

Bell Communications Research, an organization that contains much of the former Bell Labs. It specializes in telephone network technology, standards, and interfaces. Now known as Telcordia.

BER

Bit Error Rate, a measure of the quality of a data link.

BER can also refer to basic encoding rules, a part of the *ASN.1* standard.

Berkeley

Usually refers to *Unix* networking services such as *rlogin* and *sendmail* designed at the University of California at Berkeley.

Beta test

Commonly used term to describe the state of a system that is believed to be mainly functional but not yet completely tested. Beta testing is often performed by a set of trusted users or customers who are willing to accept the existence of problems and report them before the system is released for full field use. Beta testing is conventionally preceded by alpha testing, which generally indicates a system in a "just about working, many known problems, use at your own risk" state.

BHCA

Busy hour calling attempt, a measure of the throughput of a switch (typically a voice switch carrying telephony traffic).

B-ICI

Broadband Inter-Carrier Interface, a standard developed by the ATM forum from B-ISUP.

Big-endian

In terms of hardware, a computer architecture in which, within a given multibyte numeric representation, the most significant byte has the lowest address (the word is stored "big-end-first"). Most processors, including the IBM 370 family, the PDP-10, the Motorola microprocessor families, and most of the various *RISC* designs are big-endian.

Bigfoot

One of a number of global directories available on the World Wide Web. Bigfoot claims it has the most e-mail addresses available in one place—it has over three million.

BIND

Berkley Internet Name Domain. The BIND software that is used on DNS servers allows the administrator to allocate the same name to a range of Internet addresses for session balancing.

Binding

This is the process whereby a procedure call is linked with a procedure address or a client is linked to a server. In traditional programming languages, procedure calls are assigned an address when the program is compiled and linked. This is static binding. With late, or dynamic, binding, the caller and the callee are matched at the time the program is executed.

BinHex

A *Macintosh* format for representing a binary file using only printable characters. The file is converted to lines of letters, numbers, and punctuation. Because BinHex files are simply text, they can be sent through most electronic mail systems and stored on most computers. However, the conversion to text makes the file larger, so it takes longer to transmit a file in BinHex format than if the file were represented some other way. The suffix “.hqx” usually indicates a BinHex format file.

BIOS

An acronym for Basic Input/Output System. This is the lowest level interface to all peripheral devices and is usually an EPROM with computer program instructions in it. A motherboard BIOS (usually provided by companies such as Phoenix, Award, and AMI) controls the basic functions of the computer (such as controlling the keyboard or monitor). With a SCSI host adapter, the BIOS is used to control SCSI hard disk drives and perform the boot function. If a host adapter does not have a BIOS, then hard disk drives controlled by that host adapter cannot be used to boot from (booting must be done from another source, such as floppy, IDE, or another SCSI host adapter with a BIOS).

The BIOS must be enabled in order to function (e.g., a host adapter with a BIOS that is disabled acts the same as a host adapter without a BIOS). The BIOS can also contain useful software utilities, such as SCSISelect utility, which can be used to change the host adapter settings, format disks, and run simple SCSI diagnostics.

In order to provide acceptable performance, software vendors directly access routines in the BIOS rather than use the higher level operating system calls.

B-ISDN

Broadband ISDN. A high-speed extension of *ISDN* which has evolved to yield the high-speed networking technology, *ATM*. The ideas developed in narrowband ISDN (variety of services over a common link, separate signaling capability) have all carried forward into B-ISDN, and hence into ATM.

Bison

A version of the Unix compiler utility, yacc, from the Free Software Foundation.

B-ISUP

The Broadband ISDN User Part (now renamed as Q.2761-4) intended for network signaling. The ITU has developed B-ISUP as part of their signaling system number 7.

Bit

Binary digit. A unit of information in a two-state digital system. It is the amount of information obtained by asking a yes-or-no question and can take on one of two values: true and false, high and low, or 0 and 1. A bit is usually said to be set if its value is true, high, or 1, and clear if its value is false, low, or 0.

Bit error rate

The percentage of bits in a transmission that are received in error.

Bitmap

A data file or structure that corresponds bit for bit with an image displayed on a screen, probably in the same format as it would be stored in the display's video memory. A bitmap is characterized by the width and height of the image.

A bitmap may represent a colored image, in which case there will be more that 1 bit for each pixel and it might be called a "pixmap."

BITNET

An academically oriented international computer network. It is readily accessible by *Internet* users via *e-mail* and it provides a large number of databases and user groups. The name derives from "because it's time" or "because it's there."

Bits per second

The basic measurement for serial data transmission capacity, abbreviated to bps. Usually has some form of modifier: Kbps is thousands of bits per second and Mbps is millions of bits per second.

BLOB

Basic (or Binary) Large Objects, usually refers to text and image fields in databases.

BNC

The standard type of connector used to link IEEE 802.3 10base-2 coaxial cable to a transceiver.

BNF

A formal metasyntax used to express context-free grammars. Once stood for Backus Normal form, but now renamed Backus-Naur form, it is one of the most commonly used metasyntactic notations for specifying the syntax of programming languages, command sets, and the like. It is widely used for describing the syntax of a language, but is seldom documented anywhere. Even so, it can be readily understood and commonly interpreted by most people.

BOC Bell operating company. A local or regional telephone company that owns and operates lines to customer locations. The BOCs were originally established by the 1982 *MFJ* that specified the terms of the *AT&T* divestiture. The BOCs include the likes of Southern Bell and New Jersey Bell, and are sometimes known as the Baby Bells.

BOCA Business Object Component Architecture. A framework developed in the Object Management Group for component based engineering.

Bonding An international standard for aggregating multiple data channels into a single logical connection. Very popular in videoconferencing applications.

Bookmark A reference to the location (usually a network address) of a document which may or may not be on the same server to which a user is connected. Most World Wide Web and Gopher clients can save a file of bookmarks to allow you to quickly locate documents to which you want to refer frequently.

Booting The process by which a computer starts and automatically loads the operating system.

BootP Bootstrap Protocol. A protocol that a network workstation uses on boot up to determine the IP address of its Ethernet interfaces.

Boot PROM Boot Programmable Read-Only Memory. The nonvolatile memory that contains information necessary for initializing a computer system. Boot PROM information can be transmitted over a network.

Border gateway protocol The protocol used in *TCP/IP* networks for routing between different *domains.*

BORSCHT Battery feed, Over-voltage protection, Ringing, Supervision, Coding, Hybrid, and Testing. These are the essential transmission functions that have to be provided by a telecommunications provider in its local network.

Bounce

A returned *e-mail* message, which is usually sent back with some information on why it could not be delivered.

BPON

Broadband over a Passive Optical Network. A means of transmitting information over local fibers. Each customer is allocated a fixed slot within the provided time division multiplex data channels carried.

bps

Bits per second, a basic measure of data transmission speed.

BRI

Basic Rate Interface, an international standard switched digital interface offering two 64 Kbps B, or bearer, channels and a 16 Kbps D, or signaling, channel to carry voice, data, or video signals. Usually associated with ISDN.

Bridge

A relatively simple *LAN* internetworking device that filters and passes data between LANs of similar type. Bridges operate on layer 2 (*MAC* layer) information and do not use any routing algorithms. A bridge cannot work between LAN segments that use different protocols.

Broadband

A transmission medium capable of supporting a wide range of frequencies or data rates. An analog broadband network would typically run from audio up to video frequencies. It can carry multiple signals by dividing the total capacity of the medium into multiple independent bandwidth channels, where each channel operates only on a specific range of frequencies. Digital broadband is differentiated simply on the grounds of speed, with 2 Mbps commonly taken as the upper threshold.

Broadcast

A message forwarded to all network destinations.

Brooks's law

The maxim that adding programmers to a project that is late will cause it to be further delayed. Brooks explains the law in his 1975 book, *The Mythical Man-Month*, which draws on his experience from the IBM OS/360 project. The law is as true in the 1990s as it was in the 1970s.

Brouter

A device that bridges some *packets* (i.e., it forwards them based on data link layer information) and routes other packets (i.e., it forwards them based on network layer information). The bridge/route decision is based on local configuration information.

Browser

A program that allows a person to read *hypertext*. The browser gives some means of viewing the contents of nodes and of navigating from one node to another. Mosaic, Lynx, and Netscape are browsers for the *World Wide Web*. They act as clients to remote servers.

BRSA

Boundary Routing System Architecture. Software algorithms and methodology that enable a router at a central node of a wide area network to perform protocol-specific routing and bridging path table management on behalf of a router at a peripheral (leaf) node. BRSA greatly simplifies the configuration of routers at leaf nodes.

BSD

Berkeley System Distribution. In the early days of the *Unix* operating system, BSD was one of the leading implementations.

B-TERM

Broadband terminal. A SONET mechanism that provides a multiplexing function of incoming DS3 (45 Mbps) or STS-1 (155 Mbps) into a higher rate STS-N output, such as an STS-3, STS-12, or STS-48.

Buffer

Storage for blocks of data. Commonly used in networks and between computer elements to accommodate differences in rates of data transfer. The dimensioning of buffers is an important part of the detailed design of communicating systems.

Bug

An error in a program or fault in equipment. The origin of the term is disputed but the first use in a computing context is often attributed to Vice Admiral Grace Murray Hopper of the U.S. Navy. In the early days of valve-based electronic computing, she found that an error was caused by a genuine bug—a moth fluttering around inside the machine.

Bundle

General term used when a variety of products or services are combined and presented as a single offering. Bundling is increasingly prevalent on the World Wide Web, with the content coming from one source, the computers it is presented on from another, and the networks it is delivered over from another.

BUNI

Broadband User-to-Network Interface. Part of the ISDN standards.

Bursty traffic

Used to describe data rates that fluctuate widely with no predictable pattern. Devices connected to a *LAN* operate by grabbing the whole *bandwidth* for a very short period of time, and this results in peaks of activity. In contrast, connections established across a *WAN* (e.g., to download a file) tend to hold a constant data transfer rate.

BUS

Broadcast and Unknown Server. BUS provides the broadcast function and resolution of unknown addresses for LAN emulation which is connection-oriented.

Byte

A collection of binary digits. On most modern computers, a byte is 8 bits and characters are usually represented in *ASCII* in the least significant 7 bits.

Some older computers had bytes of 6 or 7 bits and the bytes in the now obsolete PDP-10 machine were actually bit fields of 1 to 36 bits! These non-standard usages are now obsolete, however.

ABCDEFGHIJKLMNOPQRSTUVWXYZABCDEFGHIJKLMNOPQRSTUVWXYZABCDEFGHIJKLMNO

C A widely used programming language, C is terse, low level, and fairly permissive in terms of the software you can produce with it. It became immensely popular outside its Bell Laboratories birthplace after about 1980, primarily due to its distribution with the *Unix* operating system. It is probably the dominant language in real-time systems and microcomputer applications programming.

C has been standardized (and at the same time modified) as *ANSI* C.

C++ An object-oriented superset of the *C* language. It was developed at Bell Laboratories in the mid-1980s and forms the basis of the *Java* language.

C7 CCITT signaling system number 7. An *ITU-T* standard for the signaling protocol that is used between intelligent network nodes such as exchanges. C7 is a common-channel signaling system. This means that signaling information is passed separately from other information. It is also known as *SS7*.

Cache A small, fast memory that holds recently accessed data and is designed to speed up subsequent access to the same data. Most often applied to processor memory access, but also used for keeping a local copy of data that have been accessed over a network and are likely to be required again (e.g., several pages of *World Wide Web* information may be cached).

Caching

This is a process by which data requested by the operating system of a computer are retrieved from RAM instead of from a hard disk (or some other mass storage media). Caching algorithms will check if the requested data are in its "cache" (or RAM). The significance of this is that RAM access is an order of magnitude faster than today's mass storage devices, so the more accesses to the cache, the faster overall system performance will be.

Cache can be on the host adapter, on the motherboard (controlled by the operating system), and on the SCSI device. Operating system vendors such as Novell and Microsoft recommend that cache not be used on the host adapter because today's operating systems cannot work in conjunction with host adapters with on-board RAM. This can lead to a degradation in performance, and possible data loss.

CAD/CAM

Computer-Aided Design/Computer-Aided Manufacturing. This is a facility normally employed in engineering design applications requiring powerful graphical facilities.

CAE

Common applications environment. An *X/Open* standard for portable applications across X/Open-conforming platforms. The applications covered include operating systems, languages, network protocols, and data management.

Call

The means by which a computer program gets access to information not held locally. There are a number of strategies for this:

1. Call by name, which is usually implemented by passing a pointer to some code, which then returns the value of the argument;
2. Call by reference, where the address of an argument variable is passed to a function or procedure;

3. Call by value, where the value of the argument expression is passed.

Increasingly, calls are being carried across networks of computers. *RPCs* have been developed to support this level of distribution.

Call center

A specially equipped, usually collocated set of operators who deal with all telephone inquiries on a specific product, topic, or service that is supported by a vendor. Call centers are usually contacted via a single number (often a toll-free number) and are supported with *ACD*, *CTI*, and dialog guidance systems so that incoming traffic can be efficiently and knowledgeably dealt with.

CALS

Computer-aided acquisition and logistics support. An electronic file format used by the U.S. government for acquiring and processing technical information.

CARGOIMP

A standard developed by the international airlines for the interchange of electronic data. This is a specific instance of *EDI*.

CAROL

Internet library on the United Kingdom's top 500 companies, an easy to use one-stop page which gathers information from anywhere on the Internet. In practice, CAROL is a librarian-cum-switchboard, able to take your call and route it through for you to the requested site.

Carrier

A continuous frequency capable of being either modulated or impressed with another information-carrying signal. Carriers are generated and maintained by modems via the transmission lines of telephone companies.

Carrier alarm

The alarm condition that is raised by the transmission of a series of 32 consecutive zeros.

Carrier signal

A continuous signal of a single frequency used by a second, data-carrying signal to convey information. The way that the second signal imparts information to the carrier signal is known as modulation.

In radio communications, the two common kinds of modulation are amplitude modulation (where the carrier amplitude is varied in line with the data to be carried) and frequency modulation (where it is the frequency that varies).

Carrier signaling

A signaling technique used in a transmission system with more than one channel. The two main alternatives are in-band and out-of-band signaling.

CASE

Computer-aided software engineering. Using computers to help with the systematic analysis, design, implementation, and maintenance of software. In the mid-1980s, CASE tools were seen by many as a route around the software production bottleneck. In retrospect, they have revealed production is only one of a range of issues that need to be attended to.

CASE

Common application service element. A general presentation layer standard from the *ISO*.

CAT

Customer Acceptance Test, usually the last stage in the commissioning of a software system.

Catalogs

These are central to eBusiness. A catalog is the electronic equivalent of a shop's shelves, goods, or departments. It is the online representation of what is "for sale" (or more correctly, what is available for trading). The way in which they are constructed is explained as are the processes for keeping them "live." This can range from a set of Web pages and a simple script that allows orders to be taken, through midrange catalog products that are characterized by a predefined structure of product categories and subcategories, up to large scale corporate catalogs that are customizable and usually feature back-end integration with inventory, stock control, and ordering systems. Catalogs for buyers and sellers are different—the former is a virtual catalog through which the buyer can see competing products from a number of suppliers and the latter is a structured set of information that represents what a supplier has to sell.

CATV Community antenna television. The name that was once used for what is now commonly known as cable television.

C Band The 6/4 GHz band used by communication satellites.

CBD Component Based Development. An approach to software engineering that encourages the creation of reusable components and the production of new software systems through the assembly of preexisting components.

CBDS The Connectionless Broadband Data Service, the counterpart of SMDS defined by ETSI. It is a connectionless message-switching network that uses ATM technology.

CBR Case-Based Reasoning, an artificial intelligence technique most frequently used to provide expert guidance (e.g., to help operators in telephone selling).

CBR Constant Bit Rate. One of the traffic types catered for by ATM. CBR, as the name suggests, delivers a link for services that require a fixed capacity to operate effectively.

CCIR Former name of the ITU committee set up to look after radio communications.

CCITT Consultative Committee of International Telegraph and Telephone. The body responsible for many telecommunications standards. Now known as *ITU-T*.

cc:mail A commercially available electronic mail package (developed by Lotus Corporation) for use with Microsoft Windows.

CCR Concurrency control and recovery. An *ISO* standard in the presentation layer used for managing database information. CCR provides the means for two or more applications to carry out mutually exclusive operations on the same data.

CCTA Central Computer and Telecommunications Agency. Former name of the Government Center for Information Systems in the United Kingdom. It is responsible for promoting the use of information technology as a means of achieving greater efficiency in both business and public sectors.

CD

Compact disc. A type of computer medium that resulted from audio technology first developed by Philips in the early 1980s, now used to carry much of the music and software sold to the general public. A standard CD can hold around 650 million bytes, the equivalent of more than 400 floppy disks.

CDDI

Copper Distributed Data Interface, a version of FDDI that works over UTP or STP copper media.

CD-I

CD-Interactive, an interactive compact disc system, with video and music, developed by Philips.

CDMA

Code-division multiple access. A technique for the transmission of multiple data streams over a single channel (usually a satellite channel). It works by assigning a different code to each separate channel so that the receiver can discriminate between each one.

CDPD

Cellular digital packet data. An evolving set of standards for the transmission of digital data over links currently used for cellular phones.

CD Plus

A standard being developed by Philips and Sony (also Microsoft, Warner, and Polygram) combining audio CD with CD-ROM, to give a mix of graphics, pictures, and video (e.g., 48 min music and 200 MB data). Works in a normal CD player.

CDR

Call detail record. A general term for the accounting record produced by network switches to track call type, time, duration, facilities used, originator, and destination. CDRs are often taken off a network exchange or switch for separate processing. They are used for a range of purposes, such as customer billing, network monitoring, and capacity planning.

CDS

Cell directory service. The local directory service provided by *DCE.*

Cell

Used in data transmission to denote a fixed number of bytes of data sent together (as in *ATM*). A cell is different from a frame in that the latter can vary in length.

"Cell" is an overloaded term which can also mean a geographic segment of a mobile phone system, a location in a spreadsheet, and a storage position for a unit of information.

Cell relay

A transmission technique based on fixed-length cells. *ATM* is the best known example of this.

CELP

Code-excited linear predictive speech. One of a number of mechanisms for reducing the bandwidth required for a voice call. CELP allows recognizable speech to be transmitted using only 8 or 16 Kbps, as opposed to the "standard" 64 Kbps.

Variations on CELP have been used to transmit voice signals over as little as 2 Kbps. This level of compression is important in that it allows voice traffic to be carried over most data links. Hence, international phone calls can be made at local rates (albeit of limited quality) simply by taking advantage of the distance-independent charging adopted on the Internet.

CEN/CENELEC

The two official European bodies responsible for setting standards. They are subsets of the members of the *ISO*. The main thrust of their work is functional standards for *OSI*-related technologies.

Centrex

A service that provides all the facilities of a *PBX*, without the need for a separate one. It is usually provided by reserving part of a network provider's switch for one group of users.

Centronics

A parallel interface for printers found on many microcomputers. It is often used for the connection of peripherals such as printers.

CEPT

The European Conference of Posts and Telecommunications. An association of European *PTTs* and network operators from 18 countries. It is the sister organization of CEN/CENELEC.

CERN

The European laboratories for particle physics. Home of the *HTML* and *HTTP* concepts, which underpin the popular Mosaic and Netscape browsers.

CGI

Common gateway interface. A protocol associated with file servers for the *World Wide Web*. CGI is the logical interface between an *HTTP* server and an *application* server. It allows information to be presented to the end user via standard forms.

Channel

In general, a transmission path between two points. For instance, the *ISDN B channel* is the transmission path for voice calls.

A specific meaning of the term is that it is the basic unit of discussion on *IRC*. Once one joins a channel, everything one types is read by others on that channel. Channels can either be named with numbers or with strings that begin with a "#" sign and can have topic descriptions (e.g., #report).

Channel aggregation

Channel aggregation combines multiple physical channels into one logical channel of greater bandwidth. With ISDN connections, channel aggregation would combine the two 64-Kb B channels into a single, logical 128-Kb channel.

Channel service unit

A type of interface used to connect a terminal or computer to a digital medium in much the same way that a modem is used for connection to an analog medium.

CHAP

Challenge, Handshaking, and Authentication Protocol. One of the security mechanisms used on public Internet services (see RFC 1334).

Character set

A collection of characters that can be represented as stored or transmitted binary data. Each character in a character set is represented by a character code that is a unique bit pattern. For example, the letter C has *ASCII* code 67.

The most widely used character set is ASCII, although others (such as Kanji, used for Japanese characters) are becoming more common, while others (such as *EBCDIC*) are now less common. Most modern operating systems support multiple character sets.

Cheapernet

A whimsical but accurate term for thin-wire *Ethernet* (*10base-2*), which uses less expensive coaxial cable instead of the full-specification yellow cable.

Checksum

Redundant digits that are added to a data packet or cell that allow it to be checked for transmission errors on receipt. A checksum usually takes the form of a simple parity check.

Chicago

In addition to being the Windy City in Illinois, this was also the code name for Microsoft's Windows 95.

CHILL

CCITT high-level language. A committee-developed programming language intended for the programming of telecommunications systems. Although not widely used, CHILL is ideally suited to (and was designed for) coding systems specified using *SDL*.

CI

Configuration item. The aggregation of hardware and software that satisfies an end-user function. Systems can be viewed as a collection of CIs, each of which is kept under separate control. The U.S. military standard MIL-STD-973 takes this controlled-component view of system construction.

CICS

Customer Information Control System. A widely used system for database handling from IBM.

CIR

Committed information rate. The nominal peak data rate for a given link in a *frame relay* network. Each frame relay permanent virtual circuit is assigned a CIR of so many bits per second. This represents the average capacity that the port connection should allocate to that link and should be consistent with the expected average traffic volume between the two sites the link connects.

The CIR represents the capacity that can be routinely relied on to be available. If there is spare capacity in the network, devices can burst above it for short periods of time.

Circuit

A communications path with a specified bandwidth. It can be either switched or permanent.

Circuit switching

The plain ordinary (or old) telephone system (*POTS*) is the best known example of a circuit-switched network. Also called *connection-oriented*, it requires a dedicated communication path to be established between the sender and receiver. All information flows over the circuit established to carry the call.

This contrasts with a connectionless or packet-switched network, where no dedicated circuit is set up and information gets from A to B over whatever routes are available.

CIT (or CTI)

Computer-integrated telephony. A combination of telephone service and computer facilities. Examples of CIT are screen-based telephones and *call centers*. Specifications for the interface between the computer and the telephone service include *TAPI* and *TSAPI*.

CIX

Commercial Internet Exchange. A major hub for Internet transmission. Connected to associated networks such as AlterNet, ANSnet, CERFnet, Eunet, Janet/JIPS, NSFnet, PSInet, and SprintNet as well as LINX.

Class

In object-oriented terms, the implementation of an object type. The same relationship exists between code and its specification. It consists of a data structure and the operational *methods* that apply to each of its objects.

CLASS

Customer-calling local-area signaling system. CLASS is a method of transferring data between a local exchange and a customer's premises. It provides a means of delivering many of the special services available through digital signaling on an analog phone (e.g., call forwarding, calling party identification).

CLASS services have grown rapidly in the United States over the last decade. Adoption in Europe, however, has been much slower.

Client

A requester of a service. More precisely, a client is an entity—for example a program, process, or person—that is participating in an interaction with another entity and is taking the role of requesting (and receiving) the required service.

At a more technical level, a client is an object that is participating in an interaction with another object and is taking the role of requesting (and receiving) the required service.

Client/server

The division of an application into two parts, where one acts as the "client" (by requesting a service) and the other acts as the "server" (by providing the service).

The rationale behind client/server computing is to exploit the local desktop processing power, leaving the server to govern the centrally held information. This should not be confused with *PCs* holding their own files on a *LAN*, since here the client or PC is carrying out its own application tasks.

Clipper

An integrated circuit that carries an *algorithm* for the encryption of telephone traffic. It is the U.S. government's preferred option for implementing secure environments.

Also the name of a popular fourth generation programming language.

CLIR

Calling Line Identity Restriction. Allows ISDN customers to request that their identities (telephone numbers) are not released at any time.

CLNS

Connectionless Network Service. Packet-switched network where each packet of data is independent and contains complete address and control information; can minimize the effect of individual line failures and distribute the load more efficiently across the network.

Clock

Any of the sources of timing signals used in isochronous data transmission.

CLS

Connectionless Protocol. A communications model where there is no explicit connection between sender and receiver. The sender just transmits the messages when ready. A real world analogy would be sending a letter. The UDP protocol is an example of a connectionless computing protocol.

Cluster

A group of user terminals that are all in the same location. They access a network via a *cluster controller*.

Cluster controller

A device such as an IBM 3x74 to which older terminal devices can be connected so that they can access network services built for modern, intelligent peripherals.

CMA

Construction and maintenance agreement. An agreement on the ownership, construction, and maintenance of expensive facilities such as undersea cables. CMAs are usually between a number of national carriers, but governments may also be included.

CMIP/CMIS

Common management information protocol/common management information service. A standard developed by the *ISO* to allow networked systems to be remotely managed. CMIS defines a message set (e.g., get, set, create, delete) along with a structure and content for the messages so that they can be used to effect control across a range of systems. In concept, it is similar to *SNMP*, but more powerful (and hence more complex).

CMOL

CMIP Over Logical link control. A proprietary network management method that uses *CMIP* over the IBM logical link control. It is used for the management of mixed-media *LANs*.

CMOT

CMIP over TCP/IP. Refers to the use of *CMIP* to manage gateways in a *TCP/IP* network. Used in the management of distributed networks.

CNET

French National Centre for Telecommunications Research: Centre National d'Etudes des Telecommunication.

CO

Connection Oriented Protocol. A communications model in which both sender and receiver must explicitly establish a connection before use, and explicitly terminate it when finished. A real-world analogy would be a telephone conversation. In the computing world, an example of a connection-oriented protocol is TCP.

COBOL

Common business-oriented language. A language for simple computations on large amounts of data. It was designed in the early 1960s, but despite its age, it is still the most widely used programming language today. COBOL has a natural language style and is intended to be largely self-documenting.

Code

A computer program expressed in a language that can be understood by the computer on which it will be executed. Source code is the human-readable (high-level language) version, and object code is the computer-readable (machine code) version. The translation from source to object is effected by a compiler or interpreter.

Codec

Encode/decode. The opposite of a modem. It converts analog signals into a form for transmission on digital circuits. The signals are then decoded back into analog at the receiving end of the transmission link. Codecs allow analog sources—typically voice and video—to be transmitted over digital links.

Collapsed backbone

Network architecture under which the backplane of a device such as a hub performs the function of a network backbone; the backplane routes traffic between desktop nodes and between other hubs serving multiple LANs.

COM

Common object model. The expansion of the component object model (see below) to add support for distribution. Developed by Digital and Microsoft, it allows interoperation between ObjectBroker and *OLE*.

COM

Component object model. The nondistributed framework underlying Microsoft's *OLE* object technology.

Common carrier

A company that offers telecommunications services to the public as well as to other carriers. Common carriers are used to provide the infrastructure for many enterprise networks.

Common-channel signaling

A scheme whereby signaling information is logically (and possibly physically) separate from the bearer circuits to which it relates. This gives a high degree of flexibility, resilience, and performance.

ISDN, with its separate *D channel* for signaling, exemplifies the common-channel idea. The term "out-of-band signaling" can also be used.

Communications controller A computer dedicated to handling the information flow to and from a number of remote terminals.

Compiler A program that converts another program from some source language (or programming language) to machine language (object code). Some compilers output assembly language which is then converted to machine language by a separate assembler. A compiler is distinguished from an assembler by the fact that each input statement does not, in general, correspond to a single machine instruction or fixed sequence of instructions.

An optimizing compiler is a compiler that attempts to analyze and improve the code it produces. A native compiler runs on the computer for which it is producing *machine code*, whereas a cross-compiler produces code for a different computer.

Components Self-contained, recognizable entities that perform well-understood functions and can be assembled via known interfaces with other components to build something more complex. Components are often reused and can be replaced with an identical functioning component without effecting overall operation.

Components Parts of a larger entity—a program, a network, or a computer system. A controlled component is sometimes known as a *CI*.

Compressed SLIP A version of *SLIP* with the *TCP* header reduced from 40 bytes down to 7.

Compression The coding of data to save storage space (and hence transmission time). Commonly available compression utilities include pack, *zip*, and pkzip.

Computer A piece of hardware that can store and execute instructions (i.e., interpret them and cause some action to occur).

Concentration

This takes consolidation a stage further by only allocating capacity when required for a customer's application such as on a call-by-call basis in the case of a telephone call.

Concentrator

Device that serves as a wiring hub in star-topology network. Sometimes refers to a device containing multiple modules of network equipment.

Concurrency

Describes a situation in which two things are happening at the same time. It is usually taken as a synonym for parallelism. There is a difference, though. With parallelism the various strands of activity are related and will eventually be reunited. This is not necessarily so with concurrency.

Configuration

A collection of items that bear a particular relationship to each other (e.g., the data configuration of a system in which classes of data and their relationships are defined). A specific software configuration is sometimes referred to as a "build."

Configuration management

An important issue to be addressed in both software development and network management. It covers the relative arrangement, connections, or options of a system from all of the available subcomponent parts (*CIs*) and objects.

Configuration management provides a basis for systems to be built from the right versions of the right components. It also allows changes to be controlled, traced, and audited.

Conformance

Meeting standards. Conformance is usually measured by running a standard set of test scripts on the product under assessment.

Connection

A communication path between two points in a network. Just as with a *circuit*, it may be either permanent or switched.

Connectionless

Communication in which two or more hosts do not have to set up a dedicated path between them before exchanging information. *Packets* are sent between any two parties, each with enough information to find their own way (perhaps over different routes) to their intended destination.

This is also called *packet switching* and is exemplified in *IP*.

Connection-oriented

In this type of connection, communication proceeds through three well-defined phases: (1) connection establishment, (2) followed by data transfer, and then (3) connection release.

The most common examples of connection-oriented communications are a telephone call and *TCP*.

CONS

Connection-Oriented Network Service. An OSI protocol for packet-switched networks that exchange information over a virtual circuit (a logical circuit where connection methods and protocols are pre-established); address information is exchanged only once. CONS must detect a virtual circuit between the sending and receiving systems before it can send packets.

Consolidation

Extraction of in-service telephony circuits from a larger number of in-service plus spare circuits. Consolidation enables the proportion of spare circuits to differ across the access network, and allows only working circuits to be connected across to core network equipment, thus avoiding unnecessarily high levels of spare stock.

Conversational

A conversational interaction is a dialog between two parties in which each "speaks" alternately.

Conway's law

States that the organization used to produce a system and structure of the system are congruent. For example, if you have three groups working on a compiler project, then they will produce a three-pass compiler.

Cookie

A token of agreement between cooperating programs that is used to keep track of a transaction. At a more concrete level, a cookie is a fragment of code that holds some information about your local state—your phone number or home page reference, for instance. You probably have cookies that you don't know about. The Netscape and Explorer browsers both support them, with the cookie being presented to the server to control your dialogue.

Cooperative multitasking

A form of multitasking in which it is the responsibility of the currently running task to give up the processor to allow other tasks to run. This contrasts with preemptive multitasking, in which there is a task scheduler that can periodically suspend running tasks and start others.

Cooperative multitasking is used in both Microsoft Windows and Macintosh System 7. It has the merit of being predictable, with tasking controlled from within a program rather than externally by a scheduler.

CORBA

Common object request broker architecture. Strictly speaking, the name of a framework specification produced by the *OMG* describing the main components of a distributed "object environment."

More loosely, CORBA is the name of any of a number of related specifications produced by the OMG that aim to provide a common approach to systems interworking.

Core

A computer's main storage. The term dates from the days of ferrite-core memory, which is now archaic.

CORE

Controlled Requirements Expression is a method for analyzing and specifying requirements.

Core dump

A copy of the contents of a computer's core, usually produced when a process is aborted by some kind of internal error.

COSE

Common Open Software Environment. An initiative by Hewlett-Packard, Sun, IBM, Novell, SCO, and other suppliers to move toward consistency and interoperability between various flavors of the *Unix* operating system.

COTS

Components Off The Shelf. The idea that you can construct a networked computing system by selecting ready made piece parts from a catalog—in much the same way that you would put together a designer bicycle from the forks, wheels, handlebars, and other parts on offer. Together with plug and play operation, COTS holds the promise of fast, customized technical solutions.

CPE

Customer premises equipment. A general term, usually used in the telecommunications community, for any piece of terminal equipment that resides with a user. Examples of CPE are phones, computers, and video terminals. The term does not usually cover larger pieces of customer equipment, such as a *PBX*.

CPI-C

Common Programming Interface for Communications. Originally an IBM standard, more portable than APPC, it has been adopted as an industry standard. Most, if not all, current implementations map to LU6.2, and can use OSI protocols.

C-plane

Control plane. A term developed as part of the ISDN structure that refers to the signaling part of the network.

CPS

Characters Per Second. A transfer rate estimated from the bit rate and length of each character. If each character is 8 bits long and includes a start and stop bit for asynchronous transmission, each character needs 10 bits to be sent. At 2,400 baud it is transmitted at approximately 240 CPS.

CPU

Central processing unit. The heart of a computer. Many CPUs are synonymous with the class of computer built around them. For instance, the 486 class of *PC* is based on an Intel 80486 processor.

CRC

Cyclic redundancy check. Additional bits appended to the end of a data stream (e.g., *packet*, *cell*) used to check whether any of the received bits are in error. The check is cyclic because the check digits are formed from the whole string as it is sent through repeated applications of an *algorithm*. The cyclic comparison of data and check digits is repeated at the receiver.

The more sophisticated the CRC, the more errors it can pick up. Simple check digits cannot usually sense errors that mask each other (e.g., two parity errors that even out). CRCs can detect this and more besides.

Although complex, there are ready-made components and algorithms for implementing many varieties of CRCs.

Creeping featurism

A common term to describe the systematic loading of more features onto a system, often at the expense of performance, maintainability, or reliability. It is symptomatic of a tendency in advanced technology for anything complicated to become even more complicated simply because technology can do it.

CRUD

Create, Read, Update, Delete. The basic functions that apply to data in a database.

Cryptography

The encoding of data so that they can only be decoded by specific individuals and hence safely and securely sent over insecure networks. It involves the design of algorithms for combining original data ("plaintext") with one or more "keys"—numbers or strings of characters usually known only to the sender and recipient—to produce coded output that should appear random to all standard statistical tests. The keys required to restore the original text may be public in some cases.

CSCW

Computer-supported cooperative work (more commonly termed *groupware*). Software tools and technology to support groups of people working together on a project, usually at different sites.

CSELT

Centro Studi E Laboratori Telecommunicazioni SpA, the Italian network operators research laboratory.

CSMA/CA

Carrier sense multiple access/collision avoidance. An arbitration protocol used on Ethernet, similar to *CSMA/CD* except that transmitters make sure that they can send without interruption. The overhead of doing this made it less efficient (and less popular) than *CSMA/CD.*

CSMA/CD

Carrier sense multiple access/collision detection. Low-level network arbitration protocol used on *Ethernet LANs.* To send information, a connected workstation must wait for quiet on the LAN before starting to transmit; it then listens during transmission.

If two workstations transmit at the same time, their data become corrupted. With CSMA/CD, they detect this and continue to transmit for a certain length of time to ensure that everyone else can detect the collision. Then they wait for a random time before attempting to transmit again, thus minimizing the chance of another collision.

CSTA

Computer-supported telecommunications application. One of a number of standards for integrating computer and telephony applications—*CTI*. CSTA is an *ECMA* standard aimed at building applications on *PBXs*.

CT2

Cordless telephone 2. A standard for cordless phones that allows users to make calls when close to a base station, as if they were using a mobile phone. CT2 phones tend to be inexpensive to own and use. CT2 is the successor to CT1 and the forerunner of the *DECT*-like CT3.

CTI (aka CIT)

Computer telephony integration. A broad term covering the integration of telephony functions with computer applications. It takes many forms in practice, from the combining of a single phone with a *PC* to the addition of a computer interface to *PBXs* and *ACDs*.

CTI is used in *call centers* to allow operators to access customer and other information while dealing with an inquiry.

Emerging standards for CTI include a set for PC-based applications (such as *TAPI* and *TSAPI*) and a set for switch equipment (such as *CSTA* and *SCAI*).

CUG

Closed User Group, a segmentation of a network that provides a select number of users with privacy within their specified group.

CU-SeeMe

An *Internet* application that enables suitably equipped users to see each other. It is interesting in that it allows visual communications (traditionally expensive and bandwidth-hungry) over standard

Internet links. Picture quality is usually far from ideal, but the ability to have pictures sent over the same link for e-mail is valuable.

Cyberspace

A term used to describe the world of computers and the society that gathers around them. First coined by William Gibson in his novel *Neuromancer.* A follow-on term—Cyberia—is sometimes used to denote an isolated existence in which the screen is a user's main outlet (it is also a cafe in London, where coffee and the Internet mix).

Cycle

The basic unit of computation. To a very rough approximation, the more cycles a computer spends working on user A's program rather than user B's, the faster user A's will run.

ABCDEFGHIJKLMNOPQRSTUVWXYZABCDEFGHIJKLMNOPQRSTUVWXYZABCDEFGHIJKLMNO

D2-MAC

A TV standard designed to replace PAL. One of the MAC family, it offers digital stereo, multilingual soundtracks, and widescreen 16:9 display (but only 625 lines), delivered by satellite. Used in continental Europe for most MAC transmissions.

DAC

Dual Attached Concentrator. A device that is attached to and allows access to both rings in an FDDI network. Also used to allude to a digital-to-analog converter.

DACS

Digital Access Carrier System, a digital pair-gain system that provides two circuits over one pair. Transmission used is a single, high-rate baseband signal. Up to 4 km.

Daemon

A program that lies dormant, waking up at regular intervals or waiting for some predetermined condition to occur before performing its action. It is supposedly an acronym derived from "disk and execution monitor." Unix systems run many daemons, chiefly to handle requests for services such as printing from hosts on a network.

DAMPS

Digital Advanced Mobile Phone Service. An AT&T product, the successor to the *AMPS* offering.

DANTE
A company established by the national research networks in Europe to provide high-capacity international network services.

DAR
Dynamic adaptive routing. The automatic rerouting of traffic based on analysis of current network loading conditions.

Dark fiber
An inactive fiber-optic strand with no associated electronics, such as transmitters, receivers, and regenerators.

DARPA
Defense Advanced Research Projects Agency. The U.S. agency that played a major part in the development of distributed systems in general and the *Internet* in particular.

DAS
Dual-Attached Station. A station with two connections to an FDDI network, one to each logical ring. If one of the rings should fail, the network automatically reconfigures to continue normal operation.

DASD
Direct-access storage device. Bulk storage for computers. It is usually referred to and pronounced as "dasdee."

DASS2
Digital access signaling system number 2. A standard used on local *ISDN* links for signaling between user and network.

DAT
Digital Audio Tape. Widely used storage media. Sampling frequencies are typically: 44.1 kHz (as used on CDs, because PCM systems used it, for compatibility with both 25-Hz PAL and 30-Hz NTSC TV signals); 48 kHz (AES/EBU standard for local transfer; professional use); 32 kHz (EBU standard for long distance transfer, not suitable for high quality music recording, but used for AM/FM radio).

Data
Numbers and characters in a form suitable for loading into a computer or transmitting over a network. The term is often used as a synonym for information, but data on their own have no meaning. Only when interpreted by some kind of data processing

system do they take on meaning and become information. For example, on its own, the number 312 is just raw data, but if it is output as your bank balance, then it becomes valuable information.

Database

A collection of interrelated data stored together with controlled redundancy to support one or more applications. On a network, data files are organized so that users can access a pool of relevant information.

There are a number of different ways to construct a database, and each option has its own characteristics. Traditional databases are built to support a relational model. Concepts here are of information stored in rows and columns with keys used to extract related items. More recently, developments have yielded object databases, where the storage item is a complete object.

Databases can be single entities, but they can also be distributed or federated. In all cases, they present a single logical image—as if the data were all in one place. The physical distribution is a design choice driven by the balance between convenience and consistency.

Database server

A stand-alone computer connected into a *LAN* that holds and manages the local database. Database management functions, such as locating the actual record being requested, are performed in the database server. The server controls access to the database itself using a *client/server* architecture. The server part of the program is responsible for updating records, ensuring that multiple access is available to authorized users, protecting the data, and communicating with other servers holding relevant data.

Data bus

The part of most computing and networking devices that makes the connections between the processor, memory, and peripherals. The data bus is often realized as a backplane bus.

Other connections are the address bus and control signals. The width of the data bus is one of the main factors determining the processing power of a computer. Most current processor designs use a

32-bit bus, meaning that 32 bits of data can be transferred at once.

Data compression

A method of reducing the amount of data to be transmitted by applying an algorithm to the basic data source. A decompression algorithm expands the data back to their original state at the other end of the link.

Compression can be "lossless" or "lossy." In the lossless case, the compression-decompression process preserves all the original information. In the lossy case, some information is sacrificed to gain greater compression. Lossy compression is typically applied to data such as digitized photographic images, where such losses are largely unimportant.

Data dictionary

A reference document developed during the design of many systems that catalogs the nature of the system data. Data dictionaries are widely used in software design and are often autogenerated as cross-references by *CASE* tools.

Data-driven

A general approach to system design that is primarily driven by data dependencies. It is generally regarded as an appropriate approach for systems that are subject to regular updates (e.g., payroll or tax calculation packages).

Data encryption key

Used for the *encryption* of message text and for the computation of message integrity checks. The latter are sometimes known as digital signatures.

Data flow

Data flow is one of a number of ways of approaching system design (others are *state* or *event driven*). With data flow, a computation is made when all the required data are available. Design focuses on this aspect, thus highlighting key data dependencies.

Datagram

A variety of data packet. A datagram is a self-contained, independent entity of data carrying enough information to be routed from source to destination without reliance on earlier exchanges between the source and destination. It is usually used to provide a one-shot message facility. An example of a datagram is the Internet-based *UDP* facility.

Data link layer

Layer 2 in the *ISO* seven-layer model. It is responsible for splitting data into packets and presenting them to the physical layer (layer 1 in the ISO model) for sending. It also checks that packets are correctly received at the end of each link. The ultimate function of the data link layer is to provide an error-free virtual channel to the network layer (layer 3 in the ISO model).

The data link layer is split into an upper sublayer, *LLC*, and a lower sublayer, *MAC*.

Data mining

Analysis of data using analytical tools that look for trends or anomalies without any specific knowledge of the meaning of the data. It is increasingly used for purposes as varied as market profiling (by examining, for instance, telephone usage patterns) or network fraud (by studying correlation between online transactions).

Data processing

Concerned with the input, verification, organization, storage, retrieval, transformation, and extraction of information from data.

The term is normally associated with commercial applications such as stock control or payroll, IBM technology, languages such as *COBOL* and *4GLs*, large databases, and transaction processing.

The data processing community is usually contrasted with the "real-time" community. The two tend to design and produce software in completely different ways, with different tools and languages.

Data transmission

The transport of digital signals. A variety of media can be used, such as optical fiber, radio, and copper wire, and speeds can range from a few hundred bits per second (using a modem) to many millions of bits per second (using *ATM*).

Datex-P

Deutsche Telekom's data services network for corporations. Datex-P has around 85% of the German market.

DAVIC

Digital Audio Visual Council. A consortium of network operators and equipment suppliers that aims to define specifications for open interfaces and protocols for all audiovisual applications and services. The remit of the DAVIC consortium includes standards for set-top boxes and other component parts of multimedia networks.

DB2

IBM's relational database that implements an SQL/DBMS product for their MVS platform.

DBMS

Database management system. A suite of programs and software tools that ease the management of large, structured sets of persistent data. DBMSs are widely used in business applications—commercial examples include Ingres, Oracle, and Sybase.

A DBMS is an extremely complex set of software programs that control the organization, storage, and retrieval of data (fields, records, and files) in a database. It usually offers ad hoc query facilities to many users. It also controls the security and integrity of the database. The DBMS accepts requests for data from the application program and instructs the operating system to transfer the appropriate data, wherever they may be physically located.

With the ever-growing amount of data being held, some form of management tool is essential. When a DBMS is used, information systems can be changed much more easily as the organization's information requirements change. New categories of data can be added to the database without disruption to the existing system.

DCE

Data communications equipment. A modem is the best known example of a DCE. In general, it is the device that links a data source (e.g., a computer) to a destination (e.g., an *ISP*).

DCE

Distributed computing environment. A set of definitions and software components for distributed computing developed by the *OSF*, an industry-led consortia.

It is primarily an *RPC* technology with integrated security and directory services, but also comprises standard programming interfaces, conventions, and server functionality (e.g., naming, distributed file system, RPC) for spreading applications transparently across networks of heterogeneous computers.

D channel

The signaling channel used in ISDN. It is used primarily for call setup and control, although it can also be used to carry packet data. Unlike the B channel, it is a permanent connection.

Originally referred to as the delta channel, the letter D was popularized because of the absence of the Greek character on word processors at that time.

DCS-1800

One of the main standards that is used in a *PCN*. The 1800 refers to the 1,800-MHz carrier band around which the standard is based.

DDE

Dynamic Data Exchange. A feature of Microsoft Windows that enables data exchange between Windows applications—the ability to exchange information, including graphical, tabular, and text, between applications (cut and paste). Superseded by OLE 2.0

DEA

Data encryption algorithm. A means of protecting data through encryption using the algorithm defined in ANSI standard X3.92. It is equivalent to the U.S. DES.

Deadlock

A condition in which two or more processes are waiting for one of the others to do something. In the meantime, nothing happens. This is an undesirable condition that needs to be guarded against, especially in the design of databases. Deadlock is also known as deadly embrace, which describes the situation wherein two or more processes are unable to proceed because each is waiting for one of the others to do something.

A common example is a workstation communicating with a server, which finds itself waiting for output from the server before it can send anything more to it. The server meanwhile is waiting for more input from the workstation before outputting anything. Sophisticated analytical design methods (such as Petri nets) have been developed to check for deadlock.

Debugging

The detection, location, and elimination of bugs. It can be a systematic exercise, but it is never complete, since the presence of a bug can be proved, but its absence cannot.

Declarative language

A general term for a functional programming language. The most common examples of declarative languages are logic programming languages like Prolog and functional languages like Haskell. These languages are well suited to some application (such as problem solving) and can be very compact, if a little terse and slow to execute.

The more widely used imperative (or procedural) languages specify explicit sequences of steps to follow to produce a result. In contrast, declarative languages describe relationships between variables in terms of inference rules, and the language executor (interpreter or compiler) applies some fixed algorithm to these relationships to produce a result.

DECnet

A very widely used proprietary network protocol designed by Digital Equipment Corporation.

DECT

Digital European cordless telecommunications. A standard for cordless connection to a variety of telecommunications networks throughout Europe (ISDN, GSM, PSTN). It is closely related to *CT2*.

Denotational semantics

A description of the meaning of a program in terms of mathematical objects. It is used primarily in the design of languages.

DES

Data encryption standard. The popular standard encryption algorithm that was originated by the U.S. National Bureau of Standards. It is intended to provide a high level of data security. DES is identical to the ANSI standard *DEA* defined in ANSI X3.92.

DES has been implemented in both software and hardware, but neither form is supposed to be distributed outside the United States. See *Clipper*.

Design

(n) A plan for a technical artifact. (v) To create a design; to plan and structure a technical artifact. In most engineering disciplines, design is the phase that is often preceded by implementation.

Design process

The process of converting a requirements specification to a set of complete manufacturing plans within the context of a chosen architecture and production environment.

Device driver

A software program that enables a PC to communicate with peripheral devices such as fixed disk drives and CD-ROM drives. Each kind of device requires a different driver. Device driver programs are stored on a PC's fixed disk and are loaded into memory at boot time.

DHCP

Dynamic host configuration protocol. A Microsoft proprietary protocol that allows IP addresses from a server to be dynamically allocated to personal computers connected on a *LAN*. It allows considerable flexibility in LAN administration.

D/I

Drop and Insert. Within a synchronous digital hierarchy, the removal (drop) or insertion (insert) of traffic at some intermediate point on an end-to-end transmission path.

Dibit

Used in *QPSK* coding, a dibit is a group of 2 bits that can be used to represent four levels—00, 01, 10, and 11.

DiffServ

Differentiated Service. A mechanism to allow IP networks to have different levels of traffic (i.e., premium, assured, and best effort). High priority traffic or premium traffic cannot be oversubscribed and this allows voice traffic to be delivered over an IP network with some assured level of quality. DiffServ uses two of the IP precedence bits to define the delivery characteristics required.

Digital certificates

These provide a means of an online retailer proving that it is truly what/whom it purports to be. The site certified by a "trusted third party" who checks the authenticity of the site and provides certified copies of that site's "public key" for communicating securely. Verisign, Inc., are world leaders in providing these services.

Digital signature

Use of a private encryption key to identify and authenticate a sender and message data.

DIN

Deutsche Industrienormen Ausshuss, the German standards-making authority, similar to the British Standards Institute.

Directory

A directory provides a means of translating from one form of information to another. In a distributed system, directory services are a key component.

Examples include *NIS*, *CDS*, and *NDS*. They usually perform much the same function as a telephone directory—translating from a symbolic name to a network address.

Directory access protocol

An *X.500* protocol used for communications between a *DUA* and a *DSA*.

Diskless workstation

A personal computer or workstation that has neither a hard disk nor floppy disk drive and that performs all of its file access via a *LAN* connection to a file server.

Distributed computing

A move away from having large centralized computers, such as minicomputer and mainframes, to bring processing power to the desktop. It is often used as a synonym for distributed processing. Distributed computing promises flexibility of system integration, but it requires a basic change from the more traditional systems philosophy (e.g., a central computer is either working or not—a distributed set of computers is usually part operational, part broken).

Distributed database

A database that allows users to gain access to records, as though they were held locally, through a database server on each of the machines holding part of the database.

Every database server needs to be able to communicate with all the others as well as be accessible to multiple users. An example is the *Internet DNS*, which provides a lookup service covering the whole network to any connected user.

Distributed processing

The distribution of information processing functions among several different locations in a distributed system.

Distributed system

A collection of computers (usually from different manufacturers) whose distribution does not concern the user. The system appears as one local machine. This contrasts with a network, where the user is aware that there are several machines and needs to know about their location, functionality, and interconnection to use them effectively. Distributed systems usually use some kind of *client/server* organization.

Dithering

Images are often made up of more colors than a particular computer can display. Many PCs, for example, can only display 256 colors on the screen. Graphic display programs (and this includes Web browsers) simulate colors that cannot be displayed directly by the process of dithering. This is achieved by creating patterns of closely spaced dots of selected colors; at a distance these combine to an approximation of the desired color. Close up the image appears to be speckled but at normal viewing distance, it is fine. Dithering is usually more visually acceptable on phtographic images as the speckling is less obvious to most people. It is less effective on block color graphics.

DLL

Data Link Layer. This is usually in two sublayers, media access control (MAC) and logical link control (LLC). MAC is the software that controls the network card, so when you buy a network card driver you're buying a MAC that accepts and presents data from the card in a known format for the higher levels.

DLL

Dynamic-link library. A library of routines that can be shared by a number of programming applications at run time. DLLs may include utility routines for standard application tasks such as managing memory and handling user actions.

DMA

Direct Memory Access. A mechanism for the hardware control of the transfer of streams of data to or from the main memory of a computing system. The mechanism may require setup by the host software. After initialization, it automatically sequences the required data transfer and provides the necessary address information.

DME Distributed Management Environment, part of DCE. Is intended to provide a system management standard for distributed systems. DME components include software license management, software distribution, event services, a subsystem management service, printing and personal computer services. Applications connect to DME using SNMP and CMIP, as well as X/Open Management Protocol (XMPAPI).

DMIG Data Management Interfaces Group. An industrial consortium set up in 1993 to develop a common set of specifications that all data management vendors agree to support, to submit to Unix vendors. DMIG includes over 25 companies (e.g., IBM, Unisys, Amdahl, Cray, Novell/Unix, SunSoft, and SCO).

DML Data Manipulation Language. The interface between an application program and a DBMS.

DNA Distributed iNternet Applications architecture. The Microsoft view of how internet-based software should operate. DNA consists of business process, storage, user identification, and navigation elements. It builds on established COM, DCOM, and Active-X ideas.

DNIC Data network identification code. A four-digit number that is assigned to public data networks to identify country and type of network. The DNIC is a part of the *ITU's X.121* standard.

DNS Domain name service. A general-purpose distributed, replicated data query service chiefly used on the Internet for translating host names into Internet addresses; for example, a dot address such as netserv.axi.com is taken and the corresponding numerical addresses are returned.

DoD The U.S. Department of Defense. Their involvement in the early days of distributed systems (primarily due to a need for resilience in the face of attack) prompted much of the subsequent development of the *Internet.*

Domain

Part of a naming hierarchy. A domain name consists, for example, of a sequence of names or other words separated by dots.

Dongle

A security or copy protection device for commercial microprocessor programs. Programs query the dongle (which needs to be inserted in the external port of the computer) before they will run.

DOS

Disk operating system. The general name for operating systems that include facilities for storing files on disk. Such a system must handle physical disk input and output, the mapping of file names to disk addresses, and the protection of files from unauthorized access (in a multiuser system). MS-DOS is probably the best known example.

DOS partition

A section of a disk storage device, created by the DOS FDISK program, in which data and/or software programs are stored. Computers have a primary DOS partition that contains the special files needed to boot the computer. A computer's disk devices may also have extended DOS partitions. Each DOS partition is assigned a unique drive letter, such as C or D. A single disk device can have multiple partitions.

Dot address

Colloquial term for an Internet address in the common dot notation (e.g., Mark@spider.ax.co.uk).

DPNSS

Digital private network signaling system. The *ISDN* signaling system used between network switches or *PBXs.* It operates at 2 Mbps and contains a sophisticated set of control commands and facilities.

DQDB

Distributed queue dual bus. An *IEEE* standard for metropolitan-area networks (e.g., one that works across LANs), the precursor to *SMDS.* DQDB was invented in Australia.

DS-0

Digital signal 0. An individual channel of 64 Kbps.

DS-1

Digital signal 1. A channel operating at 1.544 Mbps.

DS-1C

Digital signal 1C. A 3.152-Mbps transmission system.

DS-2

Digital signal 2. A 5.312-Mbps transmission system.

DS-3

Digital signal 3. A 44.738-Mbps transmission system. Same as the T3 North American standard for digital transmission service—this term is often used interchangeably with T3.

DSA

Directory systems agent. An *OSI*-defined application process that gives a user access to directory information. DSAs can be distributed to allow a user access to a remote database.

DSDM

Dynamic Systems Development Method. This is an enhancement and nonproprietary version of rapid applications development (*RAD*). DSDM defines specific roles and processes to be deployed during system development. In particular, it promotes direct user involvement, the imposition of time limits on development work, and the use of prototyping.

DSOM

An *OMG CORBA*-conformant object request broker from IBM that can be used for building distributed applications. It is related to another IBM offering, *SOM*, which provides an object environment.

DSP

Digital signal processing. Computer manipulation of analog signals (usually sounds or images) that have been converted to digital form by sampling.

DSPs are often found in the form of specialist integrated circuits and are widely used to provide processor-intensive applications such as speech guidance or image transmission. They carry out processing usually associated with software systems in circumstances where speed is of the essence.

DSS

Decision support system. Software tool, often written in a functional language, capable of deriving answers based on the input it receives plus the knowledge it contains and infers (over time) from experience.

DSSs are often used to support professional judgment; for example, the Mycin system is used to help doctors find appropriate prescriptions. Early enthusiasm for DSSs has waned with the realization that they only really work with very well-bounded subjects.

DSS1

Digital Subscriber Signaling System Number 1. The access protocol stack for basic rate ISDN. Specified by the ITU in the Q.931 standard (also relevant are Q.921, I.430/431).

DSVD

Digital Simultaneous Voice and Data. An emerging standard for the coding of signals carried over a modem. The compression algorithms used ensure that voice only takes up the smallest portion of the modem bandwidth.

DTE

Data terminal equipment. A device that acts as the source and/or destination of data and that controls the communication channel to which it is connected. A well-known example of a DTE is the personal computer. This is usually connected via an *RS-232* serial line to *DCE*, typically a modem.

DTH

Direct to home. A satellite system (also known as Death Star) that delivers a signal strong enough to be received by dish antenna a meter or less in diameter. DTH has been in service since 1994 and delivers more than 400 audio and video channels via small decoders sited by the antenna.

DTMF

Dual-tone multifrequency. A common means of signaling from a telephone to a network. Each digit dialed is represented by a pair of audio frequencies. Many network services can be invoked with this simple but effective means of signaling.

DUA

Directory user agent. An *OSI* concept that represents a user in accessing a directory. The DUA interacts with the *DSA* to obtain the services required by the user.

Dumb terminal

A terminal with no processing power of its own. Unlike an intelligent terminal, it has to be connected to a computer before it can do anything. VT100 and IBM 3270 are examples of this genre. They are usually found in large numbers attached to mainframes. This type of terminal has the virtues of being inexpensive and uncomplicated.

Duplex

A half-duplex communication channel can, at any given time, carry data in either one direction or the other, but not both. A full-duplex channel can carry data in both directions at once. A simplex channel can only ever carry data in one direction.

DVD

Digital Video Disk. A CD size disk with nearly 5-Gb capacity that can carry over two hours of high quality video and audio. All DVD players are specified to support features such as content rating and subtitling and to incorporate Dolby AC-3 digital surround sound. DVD uses MPEG-2 video compression but can also play a standard CD.

Dynamically linked library

A software library that is linked to an application when it is loaded or run rather than at the final phase of compilation. This means that the same block of library code can be shared between several tasks rather than each task containing copies of the routines it uses. The user program is compiled with a library of "stubs" that allow link errors to be detected at compile time.

Examples of operating systems using dynamic linking are SunOS, Microsoft Windows, and *RISC* OS on the Acorn Archimedes.

Dynamic binding

The property of object-oriented programming languages where the code executed to perform a given operation is determined at run time.

Dynamic routing

Routing that adjusts automatically to changes in network topology or traffic.

ABCDEFGHIJKLMNOPQRSTUVWXYZABCDEFGHIJKLMNOPQRSTUVWXYZABCDEFGHIJKLMNO

E1

A type of digital trunk circuit common throughout Europe that operates at 2.048 Mbps. E1 provides thirty 64-Kbps channels—six more than its U.S. counterpart, the T1.

E.113

ITU standard for telephone network and ISDN operation, numbering, routing and mobile service: validation procedures for the international telecommunications charge card service.

E.164

ITU standard for destination addressing scheme. Used for telephone numbers (country, area and number) as well as for SMDS data networks.

E.168

ITU standard for telephone network and ISDN operation, numbering, routing and mobile service. Specifies the application of an E.164 numbering plan for a universal personal service.

E3

A digital circuit that operates at 34 Mbps. This standard is widely used throughout Europe for intercarrier communications and is the rough equivalent of a T3.

E.410–E.414

ITU standards for network management objectives, measurements, and controls.

EBCDIC

Extended binary-coded decimal interchange code. A once popular 8-bit character encoding standard, devised by IBM, now superseded by *ASCII.* EBCDIC was an extension of the earlier 6-bit binary-coded decimal code.

eBusiness

A term that embraces all aspects of buying and selling products and services over a network. The essential characteristics of eBusiness are that the dealings between consumers and suppliers are online transactions and that the key commodity being traded is information. eBusiness is the gateway to a deal—it is a virtual entity that may (but doesn't necessarily have to) lead to physical product. Common synonyms for eBusiness include eCommerce and eTrading.

ECC

Error Correction Code, a means of detecting and correcting errors in received data. The important difference between ECC and parity checking is that ECC is capable of both detecting and correcting 1-bit errors. With ECC, 1-bit error correction usually takes place without the user even knowing an error has occurred. ECC can also detect rare 2-, 3-, or 4-bit memory errors. However, while ECC can detect these multiple-bit errors, it can only correct single-bit errors. ECC is used primarily in high-end PCs and file servers.

Echo cancellation

A process for the removal of unwanted echoes from the signal on a telephone line. These echoes are usually caused by impedance mismatches along an analog line and have to be taken out for satisfactory voice transmission and reasonable data transmission reach.

Echo cancellation is different from echo suppression, which disables the channel in one direction or the other, depending on who is talking.

ECI

European Common Interface. Part of the proposed Digital Video Broadcast set-top box (STB), it enables different service providers to use their own conditional access (CA) mechanism.

ECMA

European Computer Manufacturers Association. An association composed of members from computer manufacturers in Europe. It produces its own standards and also contributes to international bodies such as *ITU* and *ISO*.

ECSA Exchange Carriers Standards Association. An industry group responsible for the development of the original *SONET* standard.

EDI Electronic data interchange. A set of standards (*ANSI* X.12 AND *X.400*) designed to promote the exchange of orders and other business transactions by e-mail. More recently, EDI has come to be known as electronic trading.

EDIFACT EDI for Administration, Commerce, and Transport. A set of international rules for specific trading documentation such as purchase orders and invoices.

EDO Extended Data Output, or EDO. On computer systems designed to support it, EDO memory allows a CPU to access memory 10 to 15% faster than comparable fast-page mode chips.

EFF Electronic Frontier Foundation. A group established to address social and legal issues arising from the impact of the Internet and related computer-based communications networks on society. The aim of this nonprofit public interest organization is to protect freedom of expression, privacy, and access to online resources and information.

EGP Exterior gateway protocol. A TCP/IP protocol used by some routers when moving data from one system to another.

EIA Electronics Industry Association. A body that publishes standards for physical devices and their means of interfacing. *RS-232* is probably the best known EIA standard. It defines the popular 25-pin computer serial port, connector pin-outs, and electrical signaling.

EISA Extended Industry Standard Architecture. A PC bus system that is an alternative to IBM's Micro Channel Architecture (MCA). The EISA architecture, backed by an industry consortium headed by Compaq is compatible with the IBM AT bus; MCA is not.

EJB

Enterprise JavaBeans. Components written in the Java programming language intended to be run within a server based environment (e.g., a WWW-server or database). EJBs run within a "container" on the server and appear as objects to the outside world. Clients locate the EJB via the Java Naming and Directory Interface (JNDI).

Electronic data interchange (EDI)

EDI provided an early mechanism for inter-company electronic transactions. Now largely superseded by Internet-based eBusiness.

Electronic mail

Messages automatically passed from one computer user to another, often through computer networks and/or via modems over telephone lines. A message usually begins with several lines of headers giving the name and e-mail address of the sender and recipient, the time and date when it was sent, and a subject. Many other headers may get added by different message-handling systems used to move the message from sender to receiver.

The message is eventually delivered to the recipient's mailbox—a file on a computer from where it can be read using a mail-reading program such as MS-Mail, *Eudora*, or xmh.

Usually referred to as *e-mail.*

EM

Element Manager. A device that provides relatively simple management functions, such as event/alarm filtering and logging, for a specific technology (e.g., modems, T1 multiplexors).

E-mail or Email

Common shorthand for *electronic mail.*

E-mail address

The coding required to ensure that an e-mail message reaches its specified destination. There are many formats of mail address, perhaps the best known being the dot address used for Internet mail (e.g., "abe@city.ac.uk").

The *RFC* 822 standard is probably the most widely used, since it is the norm on the Internet, though *X.400* is also in use in Europe and Canada.

EMC

Electro-Magnetic Compatibility.

Emoticon

Symbols such as :-) (meaning happy) and :-((meaning sad), used to convey, when viewed sideways, an emotional state in e-mail or news. Originally intended as a joke (and known as smileys), now virtually mandatory under certain circumstances.

The reason for the widespread use of emoticons is that high-volume text-only communication forms such as Usenet have a lack of verbal and visual cues that can otherwise cause what was intended to be humorous, sarcastic, or ironic to be misinterpreted.

Encapsulation

In a data network, a method whereby one protocol is used to transport the protocol information of another. In effect, one protocol is nested inside another. It is also called “tunneling” and is used when two systems using one protocol wish to communicate over a network running another. In practice, encapsulation is something of a contingency—it does not always work and is usually inefficient.

In protocol design, it can refer to one layer adding header information to the *PDU* from the layer above. Internet protocols are a good example—a packet contains a header from the physical layer and a header added by the network layer *(IP)*, followed by a header from the transport layer (*TCP*) and then the application protocol data. Encapsulation is an important part of separating the various concerns or issues in a communications network. It allows a complex problem to be tackled one piece at a time, rather like the postal service allowing delivery (thanks to the envelope) to be handled without reference to the enclosed letter.

The term is also commonly used in software design as a means of providing users with a well-defined interface to a set of functions without revealing their internal workings.

Encapsulation is a central part of object-oriented design, where an object’s internal implementation is hidden from outside inspection or meddling. This “hiding” of internal workings is intended to help remove unnecessary complexity. Procedures that act on the objects are bundled as one unit.

Encina

Transaction processing system, from Transarc (now owned by IBM). The Unix version of CICS (CICS/6000) is based on it.

Encryption

The means of converting plaintext into ciphertext so as to prevent anyone but the intended recipient from reading the data.

Endianness

A term used to describe the packing of bytes into words within a computer processor. Processors are typically little-endian—least significant byte first—or *big-endian*—most significant first. Little-endians are typified by Intel and DEC VAX and Alpha processors.

Enterprise

A term (usually used as a descriptor for "network" or "computing") to denote the resources deployed to suit the operating needs of a particular organization.

EPROM

Erasable Programmable Read Only Memory. Chips that can be erased by placing them under an ultraviolet light for several minutes. They can then be reused.

EPS

Encapsulated PostScript (EPS) is a standard format for importing and exporting PostScript language files in all environments. It is usually a single page PostScript language program that describes an illustration. The purpose of the EPS file is to be included as an illustration in other PostScript language page descriptions. The EPS file can contain any combination of text, graphics, and images. An EPS file is the same as any other PostScript language page description, with some restrictions. EPS files can optionally contain a bitmapped image preview, so that systems that can't render PostScript directly can at least display a crude representation of what the graphic will look like. There are three preview formats: Mac (PICT), IBM (TIFF), and a platform independent preview called EPSI.

EPSF

This is a version of EPS with a TIFF preview instead of a bitmap preview.

EPSI

This is EPS with a device independent bitmap preview. EPSI is an all ASCII (no binary data or headers) version of EPS. EPSI provides for a hexadecimal encoded preview representation of the image that will be displayed or printed.

Equal access

The generic concept originating in the United States in which the *BOCs*, who control local networks, must provide access services to AT&T's competitors that are equivalent to those provided to AT&T.

Erlang

A basic unit of telecommunications traffic. One erlang is equal to 3,600 calling seconds. It is used to express the throughput of a network—the erlang rating is the ratio of the total time it is occupied to the time it is available.

ERMES

The pan-European radio-paging message system sponsored by the European Community and developed through *CEPT.* It is based on the *POCSAG* standard.

Error detection and correction

Any method of detecting and then correcting errors in transmissions. All methods involve some form of coding, the most common being the addition of a single parity bit or *CRC*, both of which are added to the end of a packet.

In general, the more extra bits that are added, the greater the chance that errors can be detected and subsequently corrected.

Error-free seconds

One way of measuring the quality of the transmitted signal. It is a percentage representation of the total amount of time (over a 24-hour period) that a signal contains bit errors. It is calculated using a test pattern defined in *ITU-T* Recommendation 0.151.

Escrow

The idea behind escrow, as applied in cryptosystems, is that a third party (e.g., a government organization) keeps private encryption keys until such time which they may use if legally empowered. The first system of this type introduced was Clipper. The introduction of this system raised considerable concern over the legal interception of communications.

ESF

Extended super frame, the current industry standard in the United States for *T1* circuits. ESF is composed of 24 frames of 192 bits each. It provides 16 signaling states in the 193rd bit for synchronization, supervisory control, and maintenance capabilities.

ESIOP

Environment-specific inter-ORB protocol. One of a number of protocol standards defined by the *OMG* for communication between *ORBs*. Others are *IIOP* and *GIOP*.

ES-IS

End system to intermediate system protocol. The OSI protocol by which end systems such as network personal computers announce themselves to intermediate systems such as hubs.

Esprit

European Strategic Program for Research in Information Technology. A funding program to develop information technology in the European economic community. Esprit worked by funding collaborations between organizations from member states across a wide range of technologies. It was superseded by the Fourth Framework.

Estelle

One of three *FDTs* (the other two being *LOTOS* and *SDL*) used within the international standards community to record specifications. Estelle has its roots in the Pascal programming language and was used mostly in France.

The original motivation for using Estelle (and the other two FDTs) was that the resulting standards would be more precise than a natural language version. In practice, the complexity of the description technique was found to confuse as much as it enlightened.

ETACS

Extended Total Access Communication System. A development of *TACS*.

Ethernet

A *LAN* characterized by 10-Mbps transmission using the *CSMA/CD* access method. Ethernet was originally developed by and is a registered trademark of Xerox Corporation. Its basic operation is that it breaks data into packets that are transmitted using the CSMA/CD algorithm until they arrive at the destination without colliding with any other.

A disk-to-Ethernet-to-disk transfer rate (with *TCP/IP*) is typically 30 Kbps. The standard Ethernet cable is 50Ω coaxial with multiple shielding.

Ethernet address

The physical address that identifies an Ethernet controller board. It is usually written as six hexadecimal numbers, for example, 38:0A:34:73:12:0C. The first three numbers identify the manufacturer of the controller.

EtherTalk

An Apple Computer network standard used to extend AppleTalk networking capability across an Ethernet network.

ETSI

European Telecommunications Standards Institute. One of a number of bodies concerned with telecommunications standards. ETSI is well known for some of its *ISDN* standards.

ETTM

Electronic Toll and Traffic Management. A system that permits cashless toll collection on roads. ETTM relies on having transponders fitted into the route. Highway 407 near Toronto is suitably equipped to run ETTM.

Eudora

E-mail software for communicating over *TCP/IP* from Macintosh, Microsoft *Windows*, Windows NT, and IBM OS/2 computers.

EURESCOM

European Institute for Research and Strategic Study in Telecommunications.

Euro-ISDN

European ISDN standard (same as I.421), issued in part by ETSI. Defines two types of services: bearer services (unrestricted 64 Kbps for voice, 3.1 kHz audio) and supplementary services (CLI, DDI, multiple subscriber number, terminal portability, CLI restriction).

Event-driven

A program that just waits for events to occur rather than running according to its own schedule. Each event has an associated handler that is passed the details of the event (e.g., user just pressed return key).

Exception handler

A special code that is called when an exception—some fault or unexpected condition—occurs. Good exception handling allows recovery or graceful collapse, as opposed to some random failure. It is an oft overlooked aspect of software design.

Exchange

The name usually used within the telecommunications community for a network switch, usually one that is used to switch *PSTN* traffic.

Modern exchanges such as GPT's System X, Ericsson's AXE 10, and Nortel's DMS 100 are large computers that carry out several stages of switching (usually time-space-time), as well as a wide range of other routing and call control functions.

Expedite

The formal process of diverging from normal protocol or procedures to accelerate the handling of a high-priority request, usually at a higher cost to the user. (The term is usually used in the context of "expedited data.")

Extranet

A term that is applied to a set of interconnected intranets. Extranets are usually established when associated organizations want to share information between themselves. In effect, they set up a closed user group by granting access privileges across each other's servers.

A B C D E F G H I J K L M N O P Q R S T U V W X Y Z A B C D E F G H I J K L M N O P Q R S T U V W X Y Z A B C D E F G H I J K L M N O

Facilities-based carrier

Carrier that uses its own facilities to provide service. It is different from a reseller—the latter purchases the services of other carriers and then retails the package of services to customers. In practice, most facilities-based carriers use some services from other carriers.

Facilities management

An organization that looks after some or all of the computing and network resources of another organization. A facilities management arrangement usually entails the provision of network and service management, rather than any actual network equipment (as would be the case with the provision of a *VPN*).

FAQ

Frequently asked questions. A set of files available over the Internet that provide a compendium of accumulated knowledge in a particular subject. FAQs tend to be maintained by volunteers. The collection of all of the FAQs is quite impressive and contains a huge wealth of up-to-date expert knowledge on many subjects of common interest, some technical, some social. As well as being posted to specific newsgroups, all FAQs are put up on the news.answers group.

Fast Ethernet

A 100-Mbps technology based on the 10base-T Ethernet CSMA/CD network access method. Competes with FDDI-over-copper.

Fast packet

A general term for a number of packet technologies with reduced overhead. It includes lightweight (and high-speed) methods such as frame relay and *ATM*. Compared with the more established *X.25* packet switching, fast packets contain a reduced functionality on the understanding that they will operate over reliable networks and will not therefore require extra baggage for error handling.

Fast SCSI

Provides for performance and compatibility enhancements to SCSI-1 by increasing the maximum synchronous data transfer rate on the SCSI bus from 5 MBps to 10 MBps. The term "fast" is generally applied to a SCSI device that can do synchronous transfers at speeds in excess of 5.0 MBps. This term can only be applied to SCSI-2 devices since SCSI-1 didn't have the timing margins that allow for fast transfers.

FAT

File allocation table. An area on a disk (hard or floppy) that acts as an index of the contents of the disk. It is used by the operating system to determine where data are located and where they can be stored.

FCAPS

Acronym for the essential elements of network management, as defined within the TMN standard. The initials stand for fault management (F), configuration management (C), accounting management (A), performance management (P), security management (S).

FCC

Federal Communications Commission. A U.S. government body overseeing and regulating national electrical and radio communications. The FCC, formed in 1934, also deals with licenses, tariffs, and limitations. The members of the commission are appointed by the U.S. president.

FCS

Frame Check Sequence (or First Customer Shipment). The former is a set of extra characters added to a frame for error control purposes. Originated with the HDLC protocol and subsequently adopted by other link layer protocols.

FDDI

Fiber distributed data interface. An *ANSI LAN* standard. It is intended to carry data between computers at speeds of up to 100 Mbps via fiber-optic links. It uses a counterrotating token ring topology and is compatible with the first (physical) level of the *ISO seven-layer model.*

FDM

Frequency-division multiplexing. The simultaneous transmission of a number of separate signals through a shared medium. The signals are split across frequency bands by modulation and routed toward a destination as a single signal before being demodulated and restored. FDM, like other multiplexing techniques, provides an economic means of transmission.

FDT

Formal description technique. A means for producing unambiguous descriptions of telecommunications services and protocols in a more precise way than can be achieved with natural language descriptions. They provide a basis for the analysis and verification of a specification. In practice, FDTs can be fairly terse, so natural language descriptions remain an essential part of a complete specification. The three FDTs in practical use are *LOTOS*, *SDL*, and *Estelle*.

Feature interaction

Knock-on effects in system design, which are most commonly experienced when an "enhancement" to one feature causes another to cease working (e.g., a feature allowing access to more databases causes the system configuration records to overwrite each other). In many ways, it is a fancy way of describing a design oversight.

FECN

Forward Explicit Congestion Notification. Used within data technologies (primarily ATM and frame relay) so that congestion is notified to the sending terminals. Not effective when congestion is of short duration, or if some terminals are unable to react.

Federation

A term often used in connection with databases or distributed computers to denote a loosely bound group of resources. In general it refers to a union of otherwise largely independent systems to support some common purpose. Federated systems share some basic agreements or protocols to enable them to work together, but are operated and managed autonomously.

Federation can provide more flexibility at the cost of greater system management and complexity.

FEP

Front-end processor. A computer that sits in front of a larger computer or system. FEPs usually perform a specific filtering task such as communications or network control, thus freeing the rest of the system to do the main task. FEPs are frequently used to control clusters of terminals accessing a mainframe.

FEPs are also becoming widely used in distributed systems to deal with processing that can be done locally, thus minimizing the delays inherent in network-based transactions. In this instance, they serve to keep processing and the data being processed close to each other.

FEPCRAPS

Fault Event Performance Configuration Resource Administration Planning and Security. Acronym for the management functions defined by Open Network Architecture.

FEXT

Far-end crosstalk. Occurs when signals from other sources leak into the input of the wrong receiver at the far end of a link. FEXT suffers more attenuation than *NEXT* and is therefore less of a problem for high-speed duplex systems.

FFOL

FDDI follow-on. The higher speed successor to FDDI, designed to operate at speeds of up to 600 Mbps. To give some idea of the impact of such speeds, the interval between bits on an E1/T1 circuit (until recently considered high-speed) is around 600 ns. At the proposed FFOL rates, this drops to 1.7 ns—enough to challenge VLSI technology simply in keeping up with the data flow.

Fiber-optic

Strands of very pure glass capable of carrying enormous volumes of traffic. Most of the high-capacity routes used by network providers use fibers.

FidoNet

An Internet-like system that comprises personal computers exchanging e-mail and files. It was originally for IBM *PCs*, but now includes *Unix* and other systems. FidoNet is a significant network with around 8,000 systems worldwide.

Fifth-generation language

Usually, programming languages that can be used to generate systems that exhibit some sort of artificial intelligence. Prolog and Lisp are often categorized as fifth-generation. It is something of a misnomer, since intelligence is not, per se, an attribute of the language.

Many millions were spent on researching fifth-generation languages in the 1980s, in the hope that the "software crisis" could be overcome this way. The cause turned out to be a lost one.

File extension

The suffix of a computer filename that tells you what sort of application was used to create that file or what sort of function that file fulfills. Common examples are:

- .com and .exe for executable files;
- .gif and .jpg for image files;
- .txt and .doc for text files;
- .bat and .sys for PC setup files.

There are a very wide variety of file extensions in use, some of which are compatible, some not.

File server

A station in a LAN dedicated to providing file and data storage to other terminals in the network. A file server is realized through a combination of hardware and software that provide file-handling and storage functions for a group of users on a LAN.

Well-known options for file server software are Sun Microsystems' Network File System, *NFS* (for Unix), and Novell *Netware* and Microsoft's LMX (for IBM PC compatibles). There is also a version of NFS for PCs that is called PC-NFS.

The advantage of file servers is that they save a network from having multiple copies of files and data stored on individual computers. As well as reducing the total disk space required, the server minimizes the problems of configuration managing a set of files shared by a number of people.

Finger

A rather quaintly named utility that helps users identify and find information on a particular user logged onto their system or a remote system. It is invoked simply by placing the system name of the user after the command "finger." It can be used to get information such as full name, last login time, alias, idle time, terminal line, and location. It is similar in function to *whois.*

Finite state machine

A way of describing a system in terms of a set of states (including the initial state), a set of input events, a set of output events, and a set of state transitions. To move through a finite state machine, the user takes the current state and an input event to determine the next state.

There is a sound theoretical basis to state machines, so they can readily be analyzed. They are well suited to the design of telecommunications and switching systems. The widely used specification language *SDL* is based on finite state machines.

Firewall

In general, this refers to the part of a system designed to isolate it from the threat of external interference (both malicious and unintentional).

Firewall code

Generally the part of a system designed to isolate it from the threat of external interference (both malicious and unintentional). For instance, there is firewall code in a telephone switch that makes sure that users cannot do any damage. Similarly, many large organizations have a firewall to protect their internal computing systems from attack.

Firewall machine

A dedicated machine that usually sits between a public network and a private one (e.g., between an organization's network and the Internet). The machine has special security precautions loaded onto it and used to filter access to and from outside network connections and dial-in lines. The general idea is to protect the more loosely administered machines from abuse.

Firmware

Software stored in read-only memory that is easier to change than hardware but less volatile than software stored on disk. Firmware is often responsible for the behavior of a system when it is first switched on.

First-generation language

Basic machine code, a means of instructing computers at the bit level.

FITL

Fiber in the Loop. A generic term for communication systems that use optical fiber rather than copper for local delivery. Passive optical networks and direct cable are both example of this.

Flag

A bit pattern consisting of six consecutive 1s sandwiched between two 0s. Flags are used in many protocols as delimiters between information fields.

Flooding

Technique where routing information received by a routing device is sent out through every interface on that device except the one on which the information was received.

FLOPS

Floating point operations per second. A measure of the speed of a processor based on the amount of useful work that it is capable of carrying out. It is often used as a benchmark.

Flow control

The variety of methods used in serial transmission to stop a sender of information from sending when the receiver is not ready to accept it.

Flow control can be implemented in both software and hardware flow control. In both cases the receiver has some sort of fixed size buffer into which received data are written when received. When the amount of buffered data hits a limit, the receiver signals to the transmitter to stop sending until the process reading the data has caught up. *X-on*, *X-off*, and the "ready to send" pin in an *RS-232* connector are common flow control mechanisms.

Formal methods

Mathematically based techniques for the specification, development, and analysis of both software and hardware systems. Examples are *Z* and *VDM*. These tend to be powerful but highly specialized and time consuming, and they tend to be used in safety-critical applications, where proof of operation cannot wait until the system goes live.

FORTRAN

Formula translator. The first and still the most widely used programming language for numerical and scientific applications. There have been a great many versions over the years. In terms of the amount of deployed software written, it is still second only to *COBOL*.

Fourth-generation language

Nonprocedural high-level languages built around database systems. Abbreviated *4GL*. 4GLs are sometimes called "program generators." In essence, they consist of an input form, screen, or template that can be fed a description of required data formats and a report types. Most 4GLs are capable of automatic code generation, usually in the form of *COBOL* programs.

4GLs can greatly speed up the production of systems, primarily those intended for data processing. Some successful examples are *SQL*, Focus, and Explorer.

FPA

Function point analysis. One of the very few standard metrics for assessing the relative size and complexity of software systems. It is usually used with systems that process a large amount of data.

Size is determined by identifying the components of the system as seen by the end user: the inputs, outputs, inquiries, interfaces to other systems, and logical internal files. The components are classified as simple, average, or complex. All of these values are then scored and weighted, and a composite complexity figure is calculated.

Like all software metrics, FPA is subject to a degree of subjective interpretation, but it has proved effective in practice.

Fractional T1

A flexible bandwidth offering that uses a portion of a 24-channel *T1* circuit. It allows a user to select anything from 2-channel 128-Kbps through all 24 channels at 1.544 Mbps in 64-Kbps increments.

Fractional T3

A service that uses a portion of a 672-channel *T3* circuit in much the same vein as fractional *T1*. Typical uses are for a mixes of voice, data, or broadcast-quality video.

FRAD

Frame relay assembler/disassembler. A device that can be used to interface a customer's *LAN* with a frame relay–based wide-area network.

Fragmentation

The breaking of a packet into smaller units when transmitting over a network medium that cannot support the original size of the packet.

Fragment units

A piece of a larger packet that has been broken down into smaller.

Frame relay

A packet-based data communications service standard that is often regarded as a lightweight version of *X.25*. It transmits bursts of data over a wide-area network in packets that vary in length from 7 to 1,024 bytes.

Frame relay is data-oriented and is typically used for LAN-to-LAN connection. Its potential latency makes it less well suited to real-time voice or video connection.

Frame relay uses the same basic framing and frame check sequence at layer 2 as at X.25, so currently installed hardware still works. It adds addressing (a 10-bit date link connection identifier, or DLCI) and a few control bits, but does not include retransmissions, link establishment, windows, or error recovery. It has none of X.25's layer 3 (session layer), but adds some simple interface management.

Frame Relay Forum

A consortium of vendors and consumers of *frame relay* equipment and services. The forum aims to specify standards that ensure interoperability between products and services from different vendors.

Freeware

Software that is provided at no charge, usually over a network such as the *Internet*. Freeware is similar to *shareware*, except that the former is given away to whomever wants it with no strings at all.

FSAN

Full Service Access Network. An industry initiative to agree on common standards for equipment that accesses a telecommunications network.

FTAM

File transfer, access, and manipulation. An *ISO* standard protocol entity forming part of the application layer, enabling users to manage and access a distributed file system.

FTP

File transfer protocol. The high-level Internet protocol for transferring files from one computer to another (it is defined in RFC 959). It is a widely used de facto standard (see the sparingly used de jure standard *FTAM*).

Anonymous FTP is a common way of allowing limited access to publicly available files via an anonymous login.

Full-Duplex Token Ring

Part of the 802.5 standard that defines dedicated and full-duplex communication for token ring networks at speeds of 32 Mbps.

Full SCSI

A SCSI solution that includes BIOS and support software to provide boot capability for hard disk and full compatibility with removable media products (e.g., hard drives, optical drives, tape drives).

Functional programming

A functional language is one kind of declarative language. Programs written in a functional language are generally compact and elegant, but have tended until recently to run slowly and require a lot of memory. Examples of functional languages are Haskell, Hope, and Miranda. Several other languages such as Lisp have a subset that is purely functional, but also contain nonfunctional constructs.

FYI

For your information. Internet bulletins that deal with topics of general interest, rather than being standard answers to specific questions. See also *FAQ* and *RFC*.

ABCDEFGHIJKLMNOPQRSTUVWXYZABCDEFGHIJKLMNOPQRSTUVWXYZABCDEFGHIJKLMNO

G.703
The *CCITT* (now ITU-T) standard for physical and logical transmissions over digital circuits. Specifications include the U.S. 1.544-Mbps version, as well as the European 2.048-Mbps version, that use the *ITU-T*-recommended physical and electrical data interface.

G.703 2.048
Transmission facilities running at 2.048 Mbps that use the *CCITT*-recommended physical and electrical data interface.

G.707
The ITU standard for SDH bit rates.

G.708
ITU-T Recommendation: Network Node Interface for SDH.

G.709
ITU-T Recommendation: Synchronous Multiplexing Structure.

G.711
An ITU standard for telephone-quality audio coding. The standard supports two (incompatible) modes: *A-law,* which is used throughout Europe, and *mu-law*, which is used in the United States and Japan. The resulting digital data rate is 64 Kbps for A-law and 56 Kbps for mu-law. In both cases, the audio bandwidth is 3.4 kHz.

Related to G.711 are G.722 (for high-quality audio coding—7-kHz bandwidth on 64/56 Kbps)

and G.728 (for low-bit audio coding—3.4 kHz on 16 Kbps).

G.722

The ITU standard for the aggregation of multiple 64-Kbps channels and the multiplexing of audio, video, and data onto the aggregated channel.

G.732

ITU standard for the characteristics of primary PCM multiplex equipment operating at 2,048 Kbps.

G.773

ITU standard for protocol suites for Q interfaces (those intended for management of transmission systems).

G.811

ITU-T Recommendation: Timing Requirements at the Outputs of Primary Reference Clocks.

Gamma

When dealing with a graphics display system, it is usually understood that when the input levels are zero (e.g., rrggbb values are zero), the output is also zero (i.e., black). Also, when the input is at its maximum (e.g., rrggbb = ffffff), the output is white. However, in between these extreme cases, there is generally some nonlinear relationship between the levels that are put into the system and those that are seen on the screen. The "gamma" of a device is a way of characterizing that nonlinearity between input and output. The formula is:

$$\text{output} = \text{input}^{\text{gamma}}$$

That is, the output is equal to the input raised to the power gamma. Here, the range of values for both input and output are chosen to be in the range 0 (black) to 1 (white). If the display system were perfectly linear, the value of gamma would be 1.0; however, display systems are not perfectly linear. A "raw" monitor has a gamma of about 2.5. Apple Macintoshes use a gamma correction in the computer which corrects it to about 1.25; most PCs, on the other hand, do not employ gamma correction and the effective gamma of the display system is thus around 2.5. Variations in gamma explain why images drawn on one computer may look quite different on another. For example, images that look good on a Mac may appear dark and dingy on a PC.

Garbage collection

The process by which storage that is no longer used is automatically reclaimed. It is widely used for programming languages that use dynamic storage allocation (such as Lisp), which are often very memory-inefficient.

Gateway

An interface between two incompatible networks. A gateway acts as a translator that enables applications to work between the two networks. It consists of hardware and software for connecting incompatible networks, which enables data to be passed from one network to another. It converts data codes and transmission protocols to enable interoperability.

GDMO

Guidelines for the Definition of Managed Objects. Provides standards and recommendations for the construction of managed objects, the basic units used to build network management systems.

General public license

The license applied to most software from the Free Software Foundation, which aims to ensure that the software is kept firmly in the public domain.

Generic markup

The technique of adding information to text that shows its contribution to the logical components of a larger document. *SGML* is an example of such a system.

GIF

Graphics Interchange Format. An image file format widely used on the Internet. More compact than the alternative JPEG (.jpg) standard but lower quality pictures. GIF files are easily spotted by their .gif extension. GIF images are compressed with an algorithm developed and owned by one of the leading online service providers, CompuServe.

GIFs support a maximum of 256 colors (i.e., 8 bits per pixel) and use the LZW compression algorithm. There are two main variants to the format: GIF87a, the form developed in 1987; and GIF89a, an extension of the standard to include transparency and animation.

Gigabit

About a thousand million bits. Or to be precise—1,073,741,824 bits.

GIGO
Garbage in, garbage out. Usually used to reinforce the fact that computers, unlike humans, will unquestioningly process the silliest of input data and quite happily produce nonsensical output.

GIOP
General inter-ORB protocol. One of a number of protocols defined by the *OMG* for communication between *ORBs*. Others are *IIOPI* and *ESIOP*.

GIS
Geographic Information System. A computer system for storing, analyzing, and displaying geographic data. Typically used for handling maps of one kind or another. GIS applications are available for navigation, recording buried plant (e.g., the location of underground cables and pipes), and many other position-dependant activities.

GKS
Graphics Kernel System. A graphical standard for 2D device-independent graphics. Specified in ISO IS7942 and ANSI X3.124-1985.

GKS-3D
Graphics Kernel System, again. A graphical standard for 3D device-independent graphics. Specified in ISO IS8805.

Glass teletype
Another term for a dumb terminal—one that has a display screen but little else.

Glue
A colloquial but very descriptive term that describes any interface logic or protocol for connecting two systems. For example, Blue Glue is IBM's *SNA* protocol.

GMT
Greenwich mean time. Otherwise known as *Zulu Time*, GMT is always the same worldwide. Communication network switches are generally coordinated using GMT as a reference.

GNN
Global Network Navigator. A collection of free services provided by O'Reilly & Associates. It includes the Whole Internet Catalog, which describes the most useful Internet resources and services; the GNN Business Pages, which list companies on the Internet; and the Internet Help Desk, which provides help in starting Internet exploration.

Gopher

One of a number of Internet-based services that provide information search and retrieval facilities. Gopher is defined in *RFC* 1432. To access Gopher, the user needs a Gopher client and needs to know the name of a Gopher server.

GOSIP

Government open systems interconnect profiles. A subset of OSI standards specific to government procurement, designed to maximize interoperability in areas where the base OSI standards are ambiguous or allow excessive options. There are GOSIPs in the United States, the United Kingdom, and other countries.

GPRS

General Packet Radio Service. A standard developed by the ETSI working groups for GSM to provide packet-orientated communications service. This will allow users to request a preferred max/min throughput and delay providing up to around 200 Kbps.

GPS

Global positioning system. A standard that allows common reference to location. It relies on a group of satellites that enable a transceiver to deduce exactly where it is and use this reference as a display on a map. It is widely used commercially for navigation and surveying.

Group III

Fax specification for transmission over analog lines at speeds up to 9,600 baud.

Group IV

Fax specification for transmission over digital (ISDN) lines. Capable of transmitting an A4 page of information in 3 sec. The three classes within the group IV specification are Class 1 for image only (200 × 200 lpi), Class 2 for images (300 × 300 lpi) plus teletex, and Class 3 for mixed mode and images up to 400 × 400 lpi.

Groupware

A general term to denote software-based tools that can be used to support a distributed set of workers. This covers applications as disparate as Windows for Workgroups and PC videophones. More formally called *CSCW*.

GSM

Global system for mobile communications. The standard for digital cellular communications that has been widely adopted across Europe and other territories. The GSM standard operates in the 900- and 1,800-MHz bands and provides a host of services, thanks to a sophisticated signaling system. It offers better speech quality than the older analog mobile standards.

Because it is digital, GSM also provides a roving connection for people to use *e-mail* and other *PC*-based facilities remotely.

GUI

Graphics user interface. An interface that enables the user to select a menu item by using a mouse to point to a graphic icon (small, simple pictorial representation of a function such as a paint brush for shading diagrams). GUI is an alternative to the more traditional character-based interface, where an alphanumeric keyboard is used to convey instructions.

This is the sort of interface most computer users have come to expect. A program with a GUI usually runs under some windowing system (e.g., the X Window System, Microsoft Windows) and uses icons, menus, and the like.

ABCDEFGHIJKLMNOPQRSTUVWXYZABCDEFGHIJKLMNOPQRSTUVWXYZABCDEFGHIJKLMNO

H.221

ITU standard for videoconferencing above 128 Kbps, for example at 256 and 384 Kbps.

H.261

A video compression standard developed by *ITU-T* primarily to support videophone and videoconferencing applications over the *ISDN*.

H.320

ITU standard that is also the industry standard for video transfer and has enabled communication between different manufacturers' products.

Half duplex

Signal flow in both directions, but only one way at a time. It is sometimes used to refer to activation of LOCAL ECHO which causes a copy of sent data to be displayed on the sending display.

Handshaking

A means of maintaining two machines or programs in synchronization. It usually requires the exchange of messages or packets of data between two systems to determine when data are sent over the link.

There are many handshaking protocols, some implemented in *hardware*, some in *software*.

Hard-coded

Data or behavior inserted directly into a program or machine, where it cannot be readily modified. This may be done for reasons of efficiency, but embedded information is a frequent source of subsequent maintenance problems.

Many systems now hold data in some profile or table that can be readily updated to accommodate a changing world.

Hardware

The physical equipment in a computer system—such as computers, network links, and *multiplexers.* Hardware used to be the dominant cost factor in systems, but has been overtaken by software cost.

Hashing

A scheme used in the design of information systems for providing rapid access to data items, which are distinguished by some *key.* Each data item to be stored is associated with a key, such as the name of a person. A hash function is applied to the item's key and the resulting hash value is used as an index for a hash table. The table contains pointers to the original items.

HCI

Human-computer interaction. Generally, the study of how humans use computers and how to design computer systems that are intuitive (and therefore easy, quick, and productive) for humans to use.

HDB3

High-density bipolar three. A line transmission standard used on *E1* circuits that eliminates data streams with eight or more consecutive zeros, thus easing potential line synchronization problems.

HDLC

High-level data link control. An *ITU-T* standard for a bit-oriented data link layer protocol. It ensures that packets are reliably transported from one point in a network to another. *HDLC* provides the link layer in *X.25* networks. Most other bit-oriented protocols are based on HDLC.

HDSL

High-bit-rate digital subscriber loop. A faster (but shorter range) variant of *ADSL.*

Header

In general networking terms, the portion of a packet, preceding the actual data, that contains source and destination addresses, error checking, and other fields.

A more specific type of header is the part of an e-mail message or news article that precedes the body of a message and contains the sender's name and *e-mail address*, the date and time the message was sent, and details of the route taken.

Heritage system

Something of a euphemism for an old system, usually one that still works but that takes up excessive or fast-growing amounts of maintenance or administrative time. The terms is synonymous with *legacy* or *stovepipe* system.

The integration of heritage systems is a headache for many organizations that have invested heavily in computer and network technology over the years. Moves toward client/server, open systems, and the like are partly motivated by a desire to minimize the impact of this problem.

Heuristic program

A system that attempts to improve its own performance as a result of learning from previous actions within a program. Such learning was one of the promised (but not delivered) innovations that was to be enabled by fifth-generation languages.

Hierarchical network

A network structure composed of layers. An example of this network can be found in a telephone network. The lower layer is the local network, which is followed by a trunk (long-distance) network up to the international exchange networks.

Hierarchical routing

Routing based on a hierarchical addressing system. IP routing algorithms use IP addresses, for example, which contain network numbers, host numbers, and, frequently, subnet numbers.

High-level language

Any computer programming language that provides some level of abstraction above assembly language. This means that it provides a set of logical statements such as "print" that correspond to several machine language instructions.

Common examples are Pascal, COBOL, *FORTRAN*, and *C*. It is much easier to program in a high-level language than in assembly language, though the speed of execution depends on how efficient the compiler or interpreter is at translating the high-level language into machine code.

HLLAPI

High-Level Language Application Program Interface. Provides a high-level interface to the 3270 control program for PC applications. HLLAPI allows PC programs to communicate directly with the mainframe.

Home page

On the *World Wide Web*, the introductory document relating to a particular site. This page often has a *URL* that is just a hostname (for instance, http://www.isoc.com, the home page for the Internet Society) and it serves to explain the structure of and provide links to underlying information.

Host

Usually refers to the large mainframe computer that a dumb terminal connects to. A host is also used for a networked computer that you can establish a session with and get some services from.

Hostage data

Data that are generally useful but held by a system that makes external access to the data difficult or expensive.

Hot Java

A *browser* from Sun Microsystems that is designed to be dynamic, so that it can incorporate new protocols as they are developed. Hot Java (formerly known as WebRunner), along with the Java language, extends *World Wide Web* documents by allowing them to include dynamic and interactive content.

Hotlink

A mechanism for sharing a piece of data between two applications. Changes made within one application (e.g., updating a spreadsheet) are reflected in the other's copy (e.g., that same spreadsheet shown as a table in a text document).

The special cut and paste commands in Windows 3 build a hotlink. *OLE* provides the more general environment of which hotlinks are one part.

Hotlist

A feature of most *World Wide Web browsers* that allows a configuration file containing hypertext links to be stored. A means of quickly reaching specific pages on the Web.

HSSI

High Speed Serial Interface. A de facto industry standard developed by a consortium including

Cisco Systems. HSSI operates over shielded cable at speeds up to 52 Mbps and distances up to 50 ft.

HTML

Hypertext Markup Language. The language used to describe the formatting in *World Wide Web* documents. It follows the *SGML* document definition where "tags" are embedded in the text to be presented. A tag consists of a symbol, a directive, zero or more parameters, and a symbol. Matched pairs of directives, like title and title are used to delimit text that is to appear in a special place or style.

As well as text layout, HTML is used to place pictures, insert buttons, and specify links to other documents. It is described in an *RFC.*

HTTP

Hypertext Transfer Protocol. The basic protocol underlying the *World Wide Web* system. It is a simple, stateless request-response protocol. Its format and use is rather similar to *SMTP.* HTTP is defined in an *RFC.*

Hub

The point where a group of circuits are connected together on a network. Hubs are used for traffic concentration and can result in economies of scale. For instance, in larger cities, hubs are deployed to concentrate and route calls from cities with lower traffic demands.

Hubs are also used in local networks. It is now common practice to organize building wiring around a hub, which can be configured (by simple patching) to provide any physical network layout as required.

Hypertext

A means of presenting documentation so that links to related text are readily apparent. Hypertext systems allow a user to click on certain words, pictures, or icons to immediately view related information for the selected item. Hypertext requires some form of language (like *HTML* or *SGML*) to specify the branches and labels that are embedded within a hypertext document.

I.113 ITU standard. Vocabulary of terms for broadband aspects of ISDN.

I.301 ITU standard. ISDN network functional principles.

I.310 ITU standard. Integrated services digital network (ISDN). Overall network aspects and functions.

I.361 ITU-T Recommendation: B-ISDN ATM layer specification.

I.420 The *ITU-T* standard that describes the options for *ISDN* Basic Rate Interface.

I.430 ITU standard. Physical layer of ISDN—part of DSS1 access protocol stack. The physical-layer specification for ISDN Basic Rate Interface service.

I.432 ITU-T Recommendation: B-ISDN UNI—physical interface specification.

IAB Internet Architecture Board. The influential panel that guides the technical standards adopted over the Internet. It is responsible for the widely accepted *TCP/IP* family of protocols. More recently, the IAB has accepted *SNMP* as its approved network management protocol.

There are two task forces under the IAB. The first is the well-known and very active *IETF* and the second is the *IRTF*.

IAB is an open body with many volunteer members from both industry and academia. IAB previously stood for Internet Activities Board.

IAYF

Information At Your Fingertips. Microsoft vision of a world where the PC is used as a general information technology tool to integrate all varieties of digital information so that it can easily be accessed without regard to type or source.

IBM PC

IBM PCs and compatible models from other vendors are the most widely used computer systems in the world. They are typically single-user personal computers, although they have been adapted into multiuser models for special applications. There are hundreds of models of IBM-compatible computers based on Intel's microprocessors: 80386, 80486, and Pentium.

All IBM personal computers are software-compatible with each other in general, but not every program will work in every machine. Some programs are time-sensitive to a particular speed class. Older programs will not take advantage of newer high-resolution display standards. The speed and generation (e.g., 486, Pentium) of the CPU (microprocessor) is the most significant factor in machine performance. This is determined by its clock rate and the number of bits it can process internally. It is also determined by the number of bits it transfers across its data bus. The second major performance factor is the speed of the hard disk.

ICMP

Internet control message protocol. The means of delivering error and control messages from hosts on the network to anyone who initiates a message. The most common use of ICMP is to test whether a destination is reachable or not. An ICMP echo is also called a "ping" (supposedly due to its similarity to the sound of a submarine's sonar). If an echo can be raised, then the host is reachable.

ICMP is defined in RFC 792.

IDE

One of a number of PC-type hard-disk interfaces.

IDL

Interface definition language. A notation that allows programs to be written for distribution across a number of processors. An IDL compiler generates stubs that provide a uniform interface to remote resources. Thus it provides a standard interface between objects within a system, and is the base mechanism for object interaction. IDL is widely used in *RPC* systems.

The Object Management Group's *CORBA* specifies the interface between objects.

IE

Information Engineering—an integrated, structured approach to developing information systems; used mainly in the business systems area for mainframe based systems.

IEC

International Electrotechnical Commission. A standardization body that is at the same level as *ISO*.

IEE

U.K. equivalent of the *IEEE*. It is a professional body covering all aspects of electronics and electrical engineering, including software, network, and computing engineering.

IEEE

Institute of Electrical and Electronics Engineers. The world's largest technical professional society, based in the United States. It was founded in 1884 by a handful of practitioners of the new electrical engineering discipline. Today's institute has more than 320,000 members.

IEEE 802

A set of IEEE standards that specify the operation of all the major aspects of *LANs*. The main items within the set are the spanning tree algorithm defined in IEEE 802.1, *LLC* (the upper portion of the data link layer) in IEEE 802.2, Ethernet in IEEE 802.3, token bus in IEEE 802.4, and IBM token ring in IEEE 802.5. The equivalent *ISO* standards are in the ISO 8802 series.

IEEE 802.3

The IEEE's specification for a physical cabling standard for *LANs*, as well as the method of transmitting data and controlling access to the cable. It uses the *CSMA/CD* access method on a bus topology LAN and is operationally similar to *Ethernet*.

IEEE 802.4

A physical layer standard that uses the *token-ring* passing access method on a bus topology *LAN*.

IEEE 802.5

A LAN physical layer standard that uses the token-ring passing access method on a ring topology *LAN*. Used by IBM on its token-ring systems.

IEEE 802.6

Metropolitan-area network standards. The basis for the DQDB architecture.

IEEE 802.9

Otherwise known as IsoEnet, a LAN standard with channels reserved for asynchronous data and for multimedia. Compatible with B-channel ISDN and therefore of interest to PBX vendors.

IEEE 802.11

Wireless Local Area Network standard.

IEEE 802.13

Fast LAN standard—100 Mbps over token ring.

IETF

Internet Engineering Task Force. A large, open international community of network designers, operators, vendors, and researchers whose purpose is to coordinate the operation, management, and evolution of the Internet and to resolve short- and mid-range protocol and architectural issues. It is a major source of proposals for protocol standards submitted to the *IAB* for final approval.

The IETF meets three times a year, and extensive minutes are included in the IETF Proceedings. The IETF Secretariat, run by the Corporation for National Research Initiatives with funding from the U.S. government, maintains an index of Internet drafts. The IAB maintains *RFCs*.

IFIP

International Federation for Information Processing. A multinational federation of professional and technical organizations (or national groupings of such organizations) concerned with information processing. From any one country, only one such organization—which must be representative of the national activities in the field of information processing—can be admitted as a full member.

IFIP was founded under the auspices of UNESCO, and there are 46 full members of the federation, representing 66 countries.

IGP

Interior gateway protocol. An extension of the *RIP*, devised by one of the leading router vendors, Cisco. It takes account of factors such as bandwidth, delay, reliability, and congestion when selecting an interconnecting link over which to send a packet.

IIOP

Internet inter-ORB protocol. One of a number of protocol standards defined by the *OMG* for communication between *ORBs*. Others are *ESIOP* and *GIOP*.

Implementation

The process of converting the notation used to express detailed software design into program code (also known as coding or programming). Implementation also denotes the task of bringing together the various system components to get the system working (also known as commissioning).

IMUX

Inverse Multiplexing. The function carried out by channel aggregators (i.e., devices for splitting a data stream into its composite 64-Kbps channels, then multiplexing back again to associate 2 or 6 of the 64-Kbps channels).

IN

Intelligent network. A fairly general term, taken to mean the addition of premium services on a public switched telephone network. The idea behind the intelligent network is that the basic network is extended in such a way that it is easy to define and implement new services. Ideally, the underlying network should not constrain the services that are added—thus allowing flexibility.

INA

Information Network Architecture. An environment developed by Bellcore into which all RBOC applications fit, replacing stand-alone operating system. Builds on distributed processing technology. The implementation of Bellcore's Open System Computing Architecture and forerunner to Telecommunications Intelligent Networking Architecture. Principally aimed at configuration and management of the information network itself.

INAP

Intelligent Network Application Protocol. Used in Intelligent Networks, as part of C7 signaling message set, for call routing.

Incremental backup

A copy of all the files that have changed (and not the ones that have remained the same) since the date of some previous backup.

Inference engine

A program that infers new facts from known facts using inference rules. It is commonly found as part of a Prolog interpreter, expert system, or knowledge-based system.

Infobahn

Information superhighway (after the German "autobahn").

Information content provider

An organization that supplies information or programming services such as news, weather, and entertainment. Such organizations are increasingly working with network and computing companies to provide online services.

Information processor

A computer-based processor for data storage and/or manipulation services for the end user.

Information retrieval

Any method or procedure used for the recovering of information or data that has been stored in an electronic medium.

Information superhighway

A much used (and often abused) term that refers to a combination of high-speed networks and sophisticated applications for information handling. Usually used in conjunction with the *Internet*, which typifies what users expect from the information superhighway.

The term originated in the United States, where the Clinton/Gore administration's plan to deregulate communication services began with their 1994 legislation to promote the integration of concepts from Internet, *CATV*, telephone providers, business networks, entertainment services, information providers, and sources of education.

Inheritance

The transfer of features (i.e., data attributes and operations) from one object to another, a concept most commonly used as part of object-oriented design. The whole idea of inheritance is that similar objects can be built or specified from a common root, thereby saving time and reducing the risk of mistakes.

Inmarsat

A set of communicating satellites. Inmarsat-A caters for maritime telephony, telex, fax and data up to 64 Kbps. Inmarsat-B is a digital version of A, including 16-Kbps voice, Group III fax, and up to 64-Kbps data. Inmarsat-C provides messaging, data reporting, electronic mail, and positioning. Inmarsat-D deals with satellite paging. Inmarsat-E is the Emergency Position Indicating Radiating Beacon (EPZRB) with low-speed messaging for position reporting. Inmarsat-Aero provides aeronautical voice and data, and Inmarsat-M, low-rate encoded voice, data, and fax for smaller sea vessels and land mobiles.

INRIA

Institut National de Recherche en Informatique et Automatique. French computer science research institute.

Integrated access

The use of a single connection to access a number of different telecommunications services (e.g., a private line, switched services, data services).

Intel 80X86

The range of processors used in the PC. Most people refer to the last three digits (e.g., 486), although the successor to the 486, the Pentium, has spoiled this convention.

Intelligent terminal

A terminal that contains a processor and memory with some level of programming facility. The opposite is a dumb terminal.

Interface

The boundary between two things, typically two programs, two pieces of hardware, a computer and its user, and a project manager and the customer.

Internet

The Internet is the largest network of computers in the world. It actually comprises many smaller networks that use the *TCP/IP* protocols to communicate and share a common addressing scheme and naming convention.

It operates over a three-level hierarchy composed of backbone networks (e.g., ARPAnet), midlevel networks, and stub networks. Logically it includes commercial (identified with a .com or .co in the address), university (.ac or .edu), military (.mil), and other research networks (.org, .net).

The Internet is recognized as the largest and most important data network in the world. It is growing at a phenomenal rate and has sparked a wealth of technical and social innovation over the years.

internet

With a lower case "i," denotes any set of networks interconnected with routers.

Internet address

The 32-bit host address defined by the *IP* in *RFC* 791. The Internet address is usually expressed in dot notation (e.g., 128.121.4.5). It can be split into a network number (or network address) and a host number unique to each host on the network and sometimes also a subnet address. The way the address is split depends on its class: A, B, or C as determined by the high address bits:

- Class A—high bit 0, 7-bit network number, 24-bit host number. n1.x.x.x, where n1 lies in the range 1 to 127.
- Class B—high 2 bits 10, 14-bit network number, 16-bit host number. n1.n2.x.x, where n1 lies in the range 128 to 191 inclusive.
- Class C—high 3 bits 110, 21-bit network number, 8-bit host number. n1.n2.n3.x, where n1 is in the range 192 to 223 inclusive.

The Internet address must be translated into an Ethernet address by either *ARP* or constant mapping.

The dramatic growth in the number of Internet users over the last few years has led to a shortage of new addresses. This is one of the issues being addressed by the introduction of a new version of *IP*, known as IPNG or *IPv6*.

Internetworking

The interconnection of two or more networks, usually local-area networks so that data can pass between hosts on the different networks as though they were one network. This requires some kind of *router* or *gateway*.

InterNIC

Internet Network Information Center. Provides a huge range of information about the Internet. It was started by the National Science Foundation, who, in cooperation with the Internet community, prompted *NIS* managers to provide and/or coordinate services for the NSFNet community.

Three organizations were selected to receive cooperative agreements in the areas of information services, directory and database services, and registration services. Together, these three awards constitute the InterNIC. General Atomics provides information services, AT&T provides directory and database services, and Network Solutions, Inc., provides registration services.

Interoperability

The ability of heterogeneous systems and networks (usually from multiple vendors) to communicate and cooperate through specified standards.

Interoperate

The ability of computers from different vendors to work together using a common set of protocols. Suns, *Vaxen*, IBMs, *Macintoshes*, *PCs*, and others all work together, allowing each to communicate with and use the resources of the other.

Interpreter

A program that translates and executes a source program one statement at a time. Interpreters are slow but easy to use.

Inverse multiplexing

A mechanism for putting lower rate data onto a higher rate bearer. An example would be using a 64-Kbps *ISDN* B channel to carry a 14.4-Kbps modem signal. Inverse multiplexing is necessary to ensure that data are correctly restored at the receiver.

The concept relates mostly to ISDN, where a considerable amount of equipment, built to send data over the analog network, can plug into an all-digital carrier. Inverse multiplexing provides the consistent padding needed to carry data of lower speed than the bearer channel.

Inverse multiplexing is the opposite of *channel aggregation*, another ISDN-related term, which uses multiple channels to carry a signal of higher speed than a single-bearer channel.

I/O

Input/output. Refers to an operations, program, or device whose purpose is to enter data into or to extract data from a computer.

IP

Internet protocol. One of the key parts of the *Internet.* IP is a connectionless, best-effort packet-switching protocol. It provides packet routing, fragmentation, and reassembly through the data link layer and supports the *TCP.* IP is defined in *RFC* 791.

IP address

Internet protocol address. A 32-bit address that has to be assigned to any computer that connects to the *Internet.* A typical IP address takes the form 192.61.33.11 and comprises a host component and a network component (see *Internet address*).

IP phone

One of a number of *internet* applications that use a standard connection to carry voice. Given the variety of internet links, the quality of transmission varies, but the fact that the same path can be used for voice as well as e-mail and file transfer is interesting.

IPSE

Integrated project support environment. A software support system that contained a range of tools—some for technical development and some for project management—all within a coherent framework.

The idea behind the IPSE was that software development could be managed in a consistent way if data integrity and development stages were closely controlled and supported.

A number of commercial IPSEs were produced in the 1980s (e.g., Genos), but they rapidly fell into disuse as the complexity of software development rendered them more of a hindrance than a help.

IP spoofing

The use of a forged IP source address to circumvent a firewall. The packet appears to have come from inside the protected network and to be eligible for forwarding into the network.

IPv6

Internet protocol version 6, also known as IPNG, or IP new generation. The most viable candidate to replace the current Internet protocol. The primary purpose of IPv6 is to solve the problem of the shortage of IP addresses. The following features have been purposed:

- A 16-byte address instead of the current 4-byte;
- Embedded encryption—a 32-bit Security Association ID (SAID) plus a variable-length initialization vector in packet headers;
- User authentication (a 32-bit SAID plus variable-length authentication data in headers);
- Autoconfiguration (currently partly handled by *DHCP*);
- Support for delay-sensitive traffic—a 24-bit flow identification field in the header to denote voice, data, or video signals being carried.

IPX

Internetwork Packet Exchange. A network protocol developed by Novell for their NetWare product for routing IP packets within building LANs.

IRC

Internet relay chat. A worldwide "party line" network that allows a user to converse with others in real time. IRC is structured as a network of Internet servers, each of which accepts connections from client programs, one per user. The IRC community and the Usenet communities overlap to some extent. Some Usenet jargon has been adopted by IRCs, as have some conventions such as *emoticons*.

IRDA

Infrared Data Association. A consortium responsible for setting standards for line-of-sight infrared protocols. This extends technology usually associated with TV remote control to speeds that are suitable for wireless data transmission between PCs and local-area networks.

IRQ

Interrupt Request Channel. The IRQ of a host adapter can be changed to several different settings by changing jumpers and/or switch settings on the adapter board.

IRTF

Internet Research Task Force. Chartered by the *IAB* to consider long-term Internet issues from a theoretical point of view. It has research groups similar to the Internet IETF working groups, which are each tasked to discuss different research topics.

Multicast audio-videoconferencing and privacy-enhanced mail are samples of IRTF output.

ISA

Industry Standard Architecture expansion bus. A type of computer bus used in most PCs. ISA enables expansion devices like network cards, video adapters, and modems to send data to and receive data from the PC's CPU and memory 16 bits at a time. Expansion devices are plugged into sockets in the PC's motherboard. ISA is sometimes called the AT bus, because it was originally introduced with the IBM PC-AT in 1983.

ISDN

Integrated services digital network. An all-digital network that allows a whole host of services to be carried together on the same circuits. It can be considered an extension of the public switched telephone network, the key similarity being that it permits any two compatible pieces of connected equipment to talk to each other. This means that ISDN can carry any form of data, such as voice, video, and computer files.

The most common ISDN system provides two high-speed (64-Kbps) data or voice circuits over a traditional copper wire pair along with a 16-Kbps signaling channel, but can scale to provide as many as 30 channels. Broadband ISDN is planned to extend the ISDN capabilities to services in the gigabit range.

ISDN currently comes in several flavors:

- A Basic Rate Interface (*BRI*) is two 64-Kbps bearer (B) channels and a single 16-Kbps signaling (D) channel.
- A Primary Rate Interface (*PRI*) in North America and Japan consists of 24 channels, usually 23 B + 1 D channels with the same physical interface as T1. Elsewhere, the PRI usually has 30 B + 1 D channels and an E1 interface.
- B-ISDN is a broadband service that is taking practical form in *ATM*.

A terminal adapter can be used to connect ISDN channels to existing interfaces such as RS-232 and V.35. ISDN is offered by local telephone companies in Australia, France, Japan, Singapore, the United Kingdom, and the United States. In some countries (e.g., Germany), the ISDN service uses a specific national signaling specification (e.g., 1.TR.6) rather than a common standard. More recently, the whole of Europe has started to phase in the *ETSI* Euro-ISDN standard.

ISDN is defined by many as the "information superhighway delivered now."

ISO

International Organization for Standardization. Commonly believed to stand for International Standards Organization. In fact, ISO is not an abbreviation—it is intended to signify commonality (derived from the Greek *iso* meaning "same"). ISO is responsible for many standards including those for data communications and computing. A well-known standard produced by ISO is the seven-layer *OSI* model.

ISO 7776

HDLC procedures. ISO 7776 is the data link layer of the common *X.25* packet-switching service.

ISO 8208

The *X.25* packet-level protocol for data terminal equipment.

ISO 8802

A set of standards that specify the operation of all of the major aspects of local-area networks. Comparable to the *IEEE 802* series.

ISO 9001

The standard most often adopted to signify software quality. In reality, it ensures no more than a basic level of process control in software development. The focus on process control (e.g., documentation, procedures) sometimes promotes bureaucracy in ISO 9001–accredited organizations.

ISOC

Internet Society. The ultimate authority for the *Internet.* ISOC is a nonprofit, professional membership organization that facilitates and supports the technical evolution of the Internet; stimulates interest in and educates the scientific and academic communities, industry, and the public about the technology and applications of the Internet; and promotes the development of new applications for the system.

The society provides a forum for discussion and collaboration on the operation and use of the global Internet infrastructure. It publishes a quarterly newsletter, the Internet Society News, and holds an annual conference.

The development of Internet technical standards takes place under the auspices of ISOC with substantial support from the Corporation for National Research Initiatives under a cooperative agreement with the U.S. federal government.

Isochronous

A form of data transmission in which individual characters are only separated by a whole number of bit-length intervals. It is typified by a transmitter that uses a synchronous clock and a receiver that does not—it detects messages by start/stop bits as in *asynchronous* transmission.

It is different from asynchronous transmission, in which the characters may be separated by random-length intervals and with synchronous transmission. An isochronous service transmits asynchronous data over a synchronous data link. An isochronous service must be able to deliver bandwidth at specific, regular intervals. It is required when time-dependent data, such as video or voice, are to be transmitted. For example, *ATM* can provide isochronous service. See also *plesiochronous*.

ISP

Internet service provider. Most people's first point of contact with the Internet. An ISP usually offers dialup access via *SLIP* or *PPP* connections to a server on the Internet. Most ISPs also offer their customers a range of client software that can be used on the net.

CompuServe and America Online are two of the best-known providers.

ISUP

ISDN user part. The higher layers of *C7*, which deals with end user signaling on the ISDN.

IT

Information technology. A general term used to cover the application of computing and communications know-how to business problems.

ITU

International Telecommunications Union. The ruling body for telecommunications and the source of many network standards.

ITU-T

International Telecommunications Union–Telecommunications. The organization responsible for making technical recommendations about telephone and data (including fax) communications systems for network operators and suppliers.

ITU-T works closely with all standards organizations to form an international uniform standards system for communication. In operation, the ITU-T splits into a number of study groups, each charged with a specific range of topics.

The responsibilities of the various study groups within the organization are:

1. Service definition;
2. Network operations;
3. Tariffs and accounting principles;
4. Network maintenance;
5. Protection against electromagnetic environment effects;
6. Outside plant;

7. Data networks and open systems communication;
8. Terminals and telematic services;
9. Television and sound transmission;
10. Languages and telecommunications applications;
11. Switching and signaling;
12. End-to-end transmission performance of networks;
13. General network aspect;
14. Modems and transmission studies for data, telegraph, and telematic services;
15. Systems and equipment for transport networks.

These study groups work toward consensus standards, and every four years the ITU as a body holds plenary sessions where it adopts new standards.

Each study group publishes its standards as a set (e.g., one of the study groups publishes the V and X series of standards and protocols: V.24 is the same as *EIA's* RS-232C).

Prior to March 1993, ITU-T was known as *CCITT*.

IVR

Interactive Voice Response. General name for systems that provide automated attendant, voice mail/messaging, voice or keyed response systems.

ABCDEFGHIJKLMNOPQRSTUVWXYZABCDEFGHIJKLMNOPQRSTUVWXYZABCDEFGHIJKLMNO

Jabber

A blanket term for a device that is behaving improperly in terms of electrical signaling on a network. With an Ethernet this is very bad, because Ethernet uses electrical signal levels to determine whether the network is available for transmission. A jabbering device can cause the entire network to halt because all other devices think it is busy.

Jabber control

The ability of a workstation to automatically interrupt transmission of abnormally long data streams (i.e., jabber frames).

JANET

Joint Academic Network. The network that links the U.K. academic and research institutes. It is an internet, with a small "i" (a large number of interconnected subnetworks), that provides connectivity within the community as well as access to external services and other communities.

JANET is based on a private X.25 packet-switched network that interconnects over 100 sites and local-area networks for end user connection. A high-speed upgrade to JANET (based on *SMDS* rather than *X.25*) called SuperJANET was introduced in the early 1990s.

Java

A compact and portable language based on C++ that has come to have significant application in the building of highly distributable applications or *applets.* Java is designed to run on a wide range of computers and to look after its own security and operation. The philosophy is that users can download anything they like over the network without having to have all of the software to use it on their local machine. With Java and applets, the idea is that the downloaded object looks after itself.

JavaBean

Components written in the Java programming language, originally intended to be delivered over the Internet and run on a desktop client PC (also see EJB).

JBIG

Joint Bi-level Image Group. A standard for the format of monochrome fax.

JCL

Job control language. The script language used to control the execution of programs in the widely used IBM OS/360's series of batch systems.

JEPI

Joint Electronic Payment Initiative. A consortium of World Wide Web suppliers aimed at allowing the separation of payment method from retailer, so your payment can be made from anywhere.

JFIF

JPEG File Interchange Format. A standard for the exchange of image files.

Jitter

The random variation in the timing of a digital signal, especially a clock. Excess jitter can make it difficult to recover data over a link.

JNDI

Java Naming and Directory Interface (see EJB).

JPEG

Joint Photographic Experts Group. The original name of the committee that designed the standard image compression algorithm. JPEG is designed to compress either full-color or gray-scale digital images of picture quality scenes.

In general, JPEG coding yields better picture quality than the comparable *GIF* coding, albeit at the cost of larger file sizes. JPEG does not work so well on nonrealistic images, such as cartoons or line drawings. An image file using JPEG can be recognized, since it uses a .jpg extension.

JSD

Jackson structured design. One of a number of popular methods for controlling the design complexity of software systems. JSD stresses the need to consider how the system under design fits into the real world. As such, it can be viewed as a precursor to the *object-oriented* approach to system development.

Jughead

A database of *gopher* links that will accept word searches and allow the results to be used on a number of remote gophers.

Junction

A transmission link that joins two exchanges.

JVM

Java Virtual Machine. The ubiquitous engine for running Java code—in essence a software CPU. The idea is that any computer can equip itself with a JVM, a small program that allows Java applets (which are widely available over the Internet) to be downloaded and used.

K — Kilobyte. A thousand bytes (actually 1,024). The term usually refers to the amount of data in a file.

KB — Kilobyte, a measure of memory capacity.

Kbps — Kilobits per second.

KDD — Kokusai Denshin Denwa, the international branch of the Japanese public network operator.

Kerberos — The name given to the security part of the *OSF DCE.* It is an authentication system, based on symmetric key cryptography, that grants users "tickets" that allow access to particular machines and services. In Greek mythology, Kerberos was a three-headed dog, who was not given to being tricked.

Kermit — A communications protocol developed to allow files to be transferred between otherwise incompatible computers. Because Kermit runs with most operating environments, it provides a near universal method for transferring files. The protocol standard is public domain, and hence Kermit ranks with *X modem* for widespread use. Kermit uses intensive encoding and error detection, and is hence fairly slow but very robust. Generally regarded as a backstop if all else fails.

Kernel

The essential part of an operating system. It is responsible for resource allocation, low-level hardware interfaces, security, and task scheduling. The kernel contains the system-level commands—those functions that are hidden from the user—and is always running on a processor.

Key

As a database concept, an identifier for a single collection of data (e.g., a particular record). The key for an employee's details might be his or her company number.

In encryption systems, a key is the digital code used with a coding algorithm to render a data stream unique once it has been encrypted. Keys can be either public or private.

Key system

A telephone system (usually one that serves an office) that provides all users with immediate access to outside lines when they press one or two special keys. Key systems usually have fewer lines and telephones than a *PBX* system. A key system can also be used with a PBX or Centrex system.

Kiosk

A public resource where one can obtain information. Multimedia kiosks are becoming increasingly popular. They provide data that are either stored locally (e.g., on CD-ROM) or accessed via a network using some kind of distributed information retrieval system such as the World Wide Web.

KIS

Knowbot Information Service (also known as netaddress). Provides a uniform user interface to a variety of remote directory services such as *whois*, *finger*, *X.500*, and MCIMail. By submitting a single query to KIS, a user can search a set of remote *white pages* and see the results of the search in a uniform format. There are several interfaces to KIS including electronic mail and *Telnet*.

Knowbot

A piece of network search software that is capable of intelligently gathering information based on a set of user-specified criteria.

Ku band

The 14/12 GHz band used by communication satellites. Ku-band antennae are typically around 2m in diameter and are often placed on the roof of a building to ensure interference-free communication channels.

ABCDEFGHIJKLMNOPQRSTUVWXYZABCDEFGHIJKLMNOPQRSTUVWXYZABCDEFGHIJKLMNO

LAN

Local-area network. A data communications network that is geographically limited (typically to a 1-km radius), allowing easy interconnection of terminals, microprocessors, and computers within adjacent buildings. *Ethernet*, *token ring*, and *FDDI* are examples of common types of LANs. Because the network is known to cover only a small area, optimizations can be made in the network signal protocols that permit data rates up to 100 Mbps.

LANE

LAN emulation. This technology provides high-speed LAN switching (i.e., bridged) interconnectivity between sites. Each LAN router will have a permanent path established to a central server. When the router has a packet to send to a destination with which it has not communicated before, it will apply to the central route server for a route to that destination. Once received, the router will forward that packet on the specified route, caching the route to be used for any subsequent packets to that destination. Also known as MPOA.

Language

An agreed-upon set of symbols, including the rules for combining them and meanings attached to the symbols, that is used to express something (e.g., the Pascal programming language, a job control language for an operating system, and a graphical language for building models of a proposed piece of software).

LAN interconnect
A point on a LAN where circuits can be routed and administered.

LAN Manager
A network operation system developed by Microsoft for PCs running IBM's OS/2, based on Intel's 80286 and 80386 microprocessors.

LAP-B
Link access protocol—balanced. The layer 2 (data link layer) protocol used by *X.25*. Technically, it is very similar to *HDLC*.

LAP-D
Link access protocol on the D channel. The *ISDN* data link protocol, used for carrying the signaling packets on the 16-Kbps channel.

LAT
A proprietary protocol for local-area networks developed by Digital. It is delay-sensitive and has been largely overtaken by other protocols, such as *IPX*.

LATA
Local Access Transport Area. A telephone company term that defines a geographic area; sometimes corresponds to an area code. A LATA is a point of access to the local network for interexchange carriers.

Latency
The time it takes for a packet to cross a network connection, from sender to receiver, or the period of time that a frame is held by a network device before it is forwarded.

Two of the most important parameters of a communications channel are its latency and its *bandwidth*.

LaTeX
One of the early document preparation systems to offer desktop publishing facilities. It used in text commands to simplify typesetting and let the user concentrate on the structure of the text rather than on formatting commands.

Layer
Communication networks for computers may be organized as a set of more or less independent protocols, each in a different layer (or level). The lowest layer governs direct host-to-host communication between the hardware at different hosts; the highest consists of user application programs.

Each layer is meant to use the layer beneath it and provide a service for the layer above. For each layer, programs at different hosts use protocols appropriate for the layer to communicate with each other. TCP/IP has five layers of protocols; OSI has seven.

The advantages of layered protocols is that the methods of passing information from one layer to another are specified clearly as part of the protocol suite, and changes within a protocol layer are prevented from affecting the other layers. This greatly simplifies the task of designing and maintaining communication programs.

Layer 1, Physical—Electrical signals and connectors;

Layer 2, Data link—Protocols and error messages;

Layer 3, Network—Addresses and routing;

Layer 4, Transport—Information exchange—delivery and flow;

Layer 5, Session—Dialog management;

Layer 6, Presentation—Mask data format differences;

Layer 7, Application—Functions and services.

In practice, the layered approach to networks is more of a conceptual aid than a practical guide. Even so, the idea is widely used and has proven invaluable in driving standards.

LCR

Least Cost Routing—a facility on many PBXs for routing calls via the most economical route automatically, even if customer dials national number.

LDAP

Lightweight Directory Access Protocol, a major component of X.500. Defines a simple means of querying data from an X.500 or any other directory. It uses many of the directory-access techniques specified in the X.500 standard, but was added to the X.500 standard because the full Directory Access Protocol was deemed to have too much code for any client with limited system resources. LDAP is more practical for mainstream usage while working over a TCP/IP link.

Lead assessor

An assessor who is qualified and is authorized to manage a quality system assessment.

Leased line

A private telephone circuit permanently connecting two points, normally provided on a lease by a local *PTT*.

Legacy system

A system that has been developed to satisfy a specific requirement and is usually difficult to substantially reconfigure without major reengineering. Such a system is also know as a stovepipe or (more charitably) a heritage system.

LEO

Low Earth orbit. An array of satellite to support mobile telephony and data services, provided by a consortia that includes Inmarsat, Iridium, Odyssey, Teledesic, and others.

Life cycle

A defined set of stages through which a development passes over time—from requirements analysis to maintenance. Common examples are the waterfall (for sequential, staged developments) and the spiral (for iterative, incremental developments). Life cycles do not map to reality too closely, but do provide some basis for measurement and hence control.

The postrelease dual of the life cycle (e.g., the phases that a product passes through as it wears out) is sometimes known as the death cycle.

Linker

A computer program that accepts the object code files of one or more separately compiled program modules and links them together into a complete executable program, resolving references from one module to another.

LINX

The London InterNet eXchange. Internet messages in the United Kingdom pass through this, for interconnection with the global Internet. LINX is the U.K. equivalent of the U.S. CIX.

LISP

List processing language. Artificial intelligence's mother tongue, a symbolic, functional, recursive language based on the ideas of lambda-calculus, variable-length lists and trees as fundamental data types, and the interpretation of code as data and vice versa. Many programmers who used LISP came to curse it as "lots of irritating superfluous parentheses."

Little-endian

A computer architecture in which, within a given 16- or 32-bit word, bytes at lower addresses have lower significance (the word is stored "little-end first"). The PDP-11 and VAX families of computers, Intel microprocessors, and a lot of communications and networking hardware are little-endian.

Livelock

A situation in which some critical stage of a task is unable to finish because its clients perpetually create more work for it to do after they have been serviced but before it can clear its queue. This differs from *deadlock* in that the process is not stopped and waiting for something else to happen, but is running flat out but can never catch up.

LLC

Logical link control. The upper portion of the data link layer, as defined in *IEEE 802.2*. The LLC sublayer presents a uniform interface to the user of the data link service, usually the network layer. Beneath the LLC sublayer is the *MAC* sublayer.

Load balancing

In routing, the ability of the router to distribute traffic over all its network ports that are the same distance from the destination address. It increases the use of network segments, which increases effective network bandwidth.

Local loop

The circuits between a telephone subscriber's residence or business and the switching equipment at the local central office.

Logic programming

A declarative, relational style of programming based on first-order logic. The original logic programming language was *Prolog*.

LOTOS

Language of Temporal Ordering Specification. A formal specification language based on temporal ordering used for protocol specification in *ISO* standards. It was published as ISO 8807 in 1990 and describes the order in which events occur.

Lotus Notes

A group of application programs from Lotus Development Corporation that allow organizations to share documents and exchange electronic mail messages. Notes supports *groupware* and has been widely used to support distributed teams.

Low-level programming

A machine-oriented language (such as *assembler*) in which each program instruction is closely linked to a single machine-code instruction.

LU

Logical Unit. One end of a communication session in an SNA network.

LU2.0

Logical Unit Type 2. An IBM term that fits within their Systems Network Architecture that provides application program communications to 3270 terminals.

LU6.2

Logical Unit Type 6.2. IBM programming interface that provides peer-to-peer communications over an SNA network. Part of APPC and also a commit protocol for transaction processing.

Lycos

One of many *World Wide Web* search facilities. Lycos is served by Carnegie Mellon University. It allows you to search document title and content for a list of keywords. Lycos is probably the biggest such index on the Web. By April 1995, the Lycos database contained 2.95 million unique documents. The Lycos database is built by a Web crawler that can bring in 5,000 documents per day. The index searches document title, headings, links, and keywords it locates in these documents. Similar search facilities are available on the World Wide Web using Alta Vista, *Yahoo*, and WebHound.

LZW

Lempel-Ziv Welch compression algorithm used for GIF images. It is "lossless" in the sense that it does not discard any image detail when compressing the image. It has been patented by Unisys.

ABCDEFGHIJKLMNOPQRSTUVWXYZABCDEFGHIJKLMNOPQRSTUVWXYZABCDEFGHIJKLMNO

M.3000

ITU standard. Principles for a Telecommunications Management Network. Comprises the initial standard for TMN and introduces basic concepts together with a number of reference configurations and standardized interfaces.

MAC

Media access control. The lower portion of the data link layer. It is the interface between a station and a network. The MAC differs for various physical media. See also *MAC address*, *Ethernet*, *LLC*, and *token ring*.

MAC address

The hardware address of a device connected to a shared network medium.

Machine code

The representation of a computer program that is actually read and interpreted by the computer. A program in machine code consists of a sequence of machine instructions (possibly interspersed with data). Instructions are binary strings that may be either all the same size (e.g., one 32-bit word for many modern *RISC* microprocessors) or of different sizes, in which case the size of the instruction is determined from the first word (e.g., Motorola 68000) or byte (e.g., Inmos transputer). The collection of all possible instructions for a particular computer is known as its instruction set.

People very rarely write programs directly in machine code. Doing so would result in speedy code, but it is tedious and error prone. Instead, they use a programming language, which is translated by the computer into machine code.

Machine-code instruction

An instruction that directly defines a particular machine operation and can be recognized and executed without any intermediate translation.

Machine learning

The ability of a machine to improve its performance based on previous results. *Neural networks* are one example of using machine learning.

Macintosh

A range of single-user, 32-bit personal computers manufactured by Apple Computer, Inc., originally based on the Motorola 68000 microprocessor family and a proprietary operating system. The Mac was notable for popularizing the graphical user interface, invented at Xerox PARC, with its easy-to-learn and easy-to-use desktop metaphor.

Although less popular than the PC, the Macintosh is seen by many as having led the way in terms of personal computing.

Macro

An executable file that stores a series of commands and/or keystrokes. It is used to invoke a frequently followed path.

MacTCP

A product providing the Macintosh with access to *TCP/IP* services.

Mailbox

(1) A file belonging to a particular user on a particular computer in which received electronic mail messages are stored, ready for the user to read them. A mailbox may be just an electronic mail address to which messages are sent and may not actually correspond to a file if the messages are processed automatically (e.g., a mail server or mailing list).

(2) In a message passing system, a mailbox is a message queue, usually stored in the memory of the processor on which the receiving process is running. Primitives are provided for sending a message to a named mailbox and for reading messages from a mailbox.

Mail bridge

A mail gateway that forwards electronic mail messages between two or more networks if they meet certain administrative criteria.

Mail gateway

A machine that connects two or more *electronic mail* systems (including dissimilar mail systems) and transfers messages between them. Sometimes the mapping and translation can be quite complex, and it generally requires a store-and-forward scheme whereby the message is received from one system completely before it is transmitted to the next system, after suitable translations.

Mail server

A program that distributes files or information in response to requests sent via electronic mail. Examples on the Internet include Almanac and Netlib. Mail servers are also used on *BITNET*.

In the days before *Internet* access was widespread and *UUCP* mail links were common, mail servers could be used to provide remote services that might now be provided via *FTP* or *World Wide Web*.

Mainframe

The traditional centralized *batch*-processing computer, placed in a computer room and serving many users. It used to be virtually synonymous with large, powerful IBM machines (such as the 370). The distinction between the mainframe and any other computer has become less clear as workstations have become more powerful.

Maintenance

Changes to a piece of software after its initial development; also called "evolution." In practice, it is the task of modifying (e.g., locating problems in, correcting, or updating) a software system after it has been put into operation.

There are several distinct types of activity that are usually called "maintenance." These are:

1. Evolution—adding new features;
2. Perfection—refining existing code to run faster, with no new functionality added;
3. Adaptation—putting existing code in a new environment (e.g., porting to a new machine);

4. Resolution—fixing faults.

Despite being little studied and frequently underplanned, maintenance accounts for about 60% of the effort invested in software systems. This is probably because maintenance is difficult to do well and is unfashionable to work on.

Majordomo

A *freeware*-mailing listserver that runs under *Unix*. It allows special-interest groups to post and receive mail to and from each other without having to continually check addresses.

Make

The *Unix* tool to automate the recompilation and linking of a set of programs, taking account of the interdependencies of modules and their modification times. Make reads instructions from a "makefile," which specifies a set of targets to build, the files on which they depend, and the commands to execute in order to produce them.

Make is a basic tool for configuration management. Most *C* systems come with a make, but it cannot, on its own, be used to control all aspects of the release of software products.

MAN

Metropolitan-area network. A data network intended to serve an area the size of a large city. MANs typically use fiber-optic cable to connect various wire LANs. Transmission speeds may vary from 2 to 100 Mbps. They are often implemented by innovative techniques, such as running optical fiber through subway tunnels. A popular example of a MAN is *SMDS*.

Manchester encoding

An error correction scheme used on Ethernet and on message pagers.

MAP

Manufacturers automation protocol. A *LAN* protocol for factory environments, developed by General Motors. MAP is a token-passing bus and is similar (but not identical) to the *IEEE 802.4* standard. The MAP protocols are designed to give predictable real-time response. MAP forms the basis of *TOP*.

MAPI

Mail Application Program Interface. Microsoft's system for sending e-mail across a *LAN*.

Marshaling

In a distributed system, the general name for the process of encoding remote procedure call data for transmission. The reverse process, converting transmitted data into local data structures, is called "unmarshaling."

MASCOT

Modular Approach to Software Construction Operation and Test. A method of software design aimed at real-time embedded systems, from the Royal Signals and Research Establishment in the United Kingdom.

MBONE

Multicast Backbone. A set of high-speed network links that provide a virtual base for multicast *IP* traffic. MBONE can be thought of as a high-capacity part of the Internet. IP multicast-based routing allows distributed applications to achieve real-time communication over IP wide-area networks through a lightweight, highly threaded model of communication.

Each network provider participant in the MBONE provides one or more IP multicast *routers* (mrouters) to connect with tunnels to other participants and to customers. The multicast routers are typically separate from a network's production routers, since most production routers do not yet support IP multicast. Most sites use workstations running the mrouted program, but the experimental *MOSPF* software for Proteon routers is an alternative. *RFC* 1112 gives the details.

MBONE is used to give public access desktop video communications. The quality is poor, with only 3 to 5 frames per second instead of the 30 frames per second of commercial television. The great advantage, though, is that it avoids all telecommunications costs normally associated with teleconferencing. An interesting innovation is the use of MBONE for audio communications and an electronic "whiteboard," where the computer screen becomes a shared workspace where two physically remote parties can draw on and edit shared documents in real time.

Mbps Megabits per second (one million bits per second).

MBWA Management by walking about. An approach taken by those who believe that one learns more about reality by talking to those at the coal face than by poring over documents such as project reports. Projects that follow this approach tend to be more successful.

MCA Micro Channel Architecture: IBM PS/2 (models 50-95) and compatible computers have an MCA computer bus inside. These can be driven by multiple independent bus master processors. MCA is the basis for the IBM Micro Channel bus used in high-end models of IBM's PS/2 series of personal computers.

MDF Main distribution frame. Provides a physical point of flexibility within a network exchange or switch. It allows incoming and outgoing lines to be patched as required.

Megabyte A million bytes, or more precisely, 1,024 kilobytes. Abbreviated MB.

MEO Medium Earth orbit. An efficient means of providing global transmission. MEO requires fewer satellites and ground stations than LEO.

Message passing Communication through the exchange of messages. Although not a rigorously used term, message passing systems usually have the connotation of real-time immediate message exchange.

Message queuing A message-passing technology augmented by a store-and-forward capability.

Message switching General term to describe data communications where a whole message is stored and then forwarded to one or more destinations when they are free to receive traffic. Because of the comparatively long transit time through networks, message switching tends to be used for administrative messages rather than data.

Method

A way of doing something. In software terms, it is generally a defined approach to achieving the various phases of the life cycle. Methods are usually regarded as functionally similar to tools (e.g., a specific tool will support a particular method).

In object-oriented systems, method is the specific set of activities that an object can carry out; in effect, it is the actions that the object is capable of.

MFJ

Modified Final Judgment. A significant milestone in the development of the U.S. telecommunications industry, the MFJ specified the divestiture of AT&T and created the seven regional *BOCs* and *equal access*. AT&T retained long-distance service and its manufacturing business. The restriction that barred AT&T from entering the computer business was lifted.

MFLOPS

Megaflops, that is, Millions of Floating Point Operations.

MHEG

Multimedia and Hypermedia Information Coding Experts Group. A body within the *ISO* that develops standards for the coding and interaction of multimedia objects.

MHS

Message-handling system. The standard defined by ITU-T as *X.400* and by *ISO* as the Message-Oriented Text Interchange Standard (MOTIS). MHS is the X.400 family of services and protocols forming part of the applications layer, which provides a generalized facility for exchanging messages between systems—the basis for global electronic mail transfer among local mail systems. It is used by CompuServe, among others.

MIB

Management information base. The data schema that defines information available from an *SNMP*-manageable device or service to network management systems. It is realized as a database of managed objects that can be accessed by network management protocols to provide a managed network service.

An SNMP MIB is a set of parameters that an SNMP management station can query or set in the

SNMP agent of a network device (e.g., router). SNMP has two standard MIBs. The first, MIB I, was established in RFC 1156 and was defined to manage TCP/IP-based internets. MIB II, defined in RFC 1213, is basically an update of MIB I.

Microkernel

An approach to the design of operating systems that puts emphasis on small modules that implement the basic features of the system kernel and can be flexibly configured.

Microsoft BackOffice

An integrated suite of client/server applications: Windows NT Server 3.5, SQL Server, SNA Server, SMS, and Mail. An open platform backed with support and service offerings from Microsoft and Solutions Providers.

Microsoft Network

An interactive global online information network. A rival to CompuServe and other online information providers. Available as a standard part of Windows.

Middleware

Software that mediates between an application program and a network. It manages the interaction between disparate applications across the heterogeneous computing platforms. The *ORB*, software that manages communication between objects, is an example of middleware.

MIDI

Musical Instrument Digital Interface. The standard interface between musical instruments and computers.

Mid-tier

The part of computing infrastructure that acts as an interface, arbiter, security monitor, gateway, or federator of information held on back-end systems. Typically, these back-end systems would be old (legacy, or cherished) computers that hold large amounts of valuable data and the mid-tier would be in place to collect, collate, and manipulate the data for presentation via a modern front end such as a browser.

MIL

Management Information Library. A database containing instances of network management information.

MIMD

Multiple Instruction Multiple Data. One of a number of ways of configuring a distributed system.

MIME

Multipurpose Internet mail extensions. A method of file identification such that the first packet of information received by a client tells it about the type of file the server has sent (e.g., PostScript, Word document).

MIME has been adopted as a standard for multipart, multimedia electronic mail messages and World Wide Web hypertext documents on the Internet. It provides the ability to transfer nontextual data, such as graphics, audio, and fax. It is defined in RFC 1341.

Miniature Card

A small form memory expansion card that supports many different markets and applications, including audio recording, digital photography, cellular phone, handheld PCs, and other small portable electronic devices. The Miniature Card is designed as a removable, reusable digital storage medium for handheld consumer electronic devices such as digital cameras, voice recorders, smart cellular phones, and personal digital assistants (PDAs). These all require small, rugged data storage media as well as a convenient method for bringing the data to a PC for manipulation. The Miniature Card specification includes: a flexible host design that can accommodate up to 64 MB of flash, ROM, and DRAM memory; a rugged pinless connector well-suited for consumer use; a 60-connection bus for ease of system integration and low-cost implementation; and a form factor measuring 38 mm (width) × 33 mm (length) × 3.5 mm (height) for minimized card footprint. A 16-bit data path for high-performance transfers with high-capacity data storage also enables application or operating system code to be directly executed from the Miniature Card.

MiniDisc

A small CD format from Sony giving up to 74 minutes playing time, using ATRAC for coding.

MIPS

Millions of instructions per second. This used to be the key parameter in the comparison of processor performance.

Mirror

(1) Writing duplicate data to more than one device in order to protect against loss of data in the event of device failure. This technique may be implemented in either hardware or software.

Interestingly, when this technique is used with tape storage systems, it is usually referred to as "twinning." A less expensive alternative, which only limits the amount of data loss, is to make regular backups from a single disk to magnetic tape.

(2) An archive site that keeps a copy of some or all files at another site so as to make them more quickly available to local users and to reduce the load on the source site. Such mirroring is usually done for specific whole directories or files on a specific remote server, as opposed to a cache or proxy server that keeps copies of everything that is requested through it.

MIS

Management information systems. Generally indicates a data-intensive system designed to distill management summaries from a welter of information. MIS is usually associated with mainframe implementations, where large amounts of data are stored and processed, but where real-time response times are not required.

MMI

Man-machine interface. A term usually superseded by *GUI* that denotes the interface between an end user and a computer.

MML

Man Machine Language. These provide a means for operators to manipulate computer systems without having to access the computer's software.

MMO

Magnetic Field Modulation Over-write: used on Sony MiniDisc for recording. With MMO, a laser heats up the underside of the terbium ferrite cobalt disc above curie temperature, then a magnetic field stores the data.

MO

Managed Objects. Term to describe the devices, systems, protocols, or applications that require some form of monitoring and management.

Model

An abstraction of reality that bears enough resemblance to the object of the model that it is possible to answer some questions about the object by consulting the model.

Modeling

Simulation of a system by manipulating a number of interactive variables. It is possible to answer "what if...?" questions to predict the behavior of the modeled system. A model of a system or subsystem is often called a "prototype."

Modem

Modulator-demodulator. Data communications equipment that performs necessary signal conversions to and from terminals to permit transmission of source data over telephone and/or data networks.

Modularization

The splitting up of a software system into a number of sections (modules) to ease design, coding, testing, and overall ease of management. It only works if the interfaces between the modules are clearly and accurately specified.

MOM

Message-oriented *middleware*. Used to describe commercial message-passing and message-queuing products.

Moore's Law

The observation that the logic density of silicon integrated circuits has closely followed the curve $d = 2$ raised to the power $(t - 1962)$, where d is the density in bits per square inch and t is the current year. It translates into the fact that the amount of information storable on a given amount of silicon has roughly doubled every year since the technology was invented.

By way of a counter, there is the associated Parkinson's Law of Data, which states that data expand to fill the space available for storage; so buying more memory encourages the use of more memory-intensive techniques.

Both of these laws have been observed to hold over the last 10 years. Unfortunately, the laws of physics guarantee that Moore's law cannot continue indefinitely.

Mosaic

NCSA's browser (client) for the *World Wide Web.* Mosaic has been described as "the killer application of the 1990s" because it was the first program to provide an intuitive, multimedia, and graphical user interface to the *Internet's* burgeoning wealth of distributed information services (formerly mostly limited to *FTP* and *gopher*) at a time when access to the Internet was expanding rapidly outside its previous domain of academia and large industrial research institutions.

Motif

The standard graphical user interface and window manager from *OSF*, running on the X Window System.

MOTIS

Message-oriented text interchange standard. A message-handling system, the ISO equivalent of the ITU X.400 standard.

Motorola 68000

The first member of Motorola's family of 16- and 32-bit microprocessors. It was the successor to the Motorola 6809 and was followed by the Motorola 68010. The 68000 was used in many workstations, notably, early Sun 2s and Sun 3s, and personal computers, notably, Apple Computer's first Macintoshes.

MP

Multiprocessing or message passing.

MPEG

Moving Picture Experts Group. An *ISO* committee that generates standards for digital video compression. It is also the name of its algorithm. MPEG1 is optimized for CD-ROM. More recent variants are MPEG2 for broadcast-quality video and MPEG4 for low-bandwidth videotelephony.

MPEG2

Moving Picture Experts Group. The group produces a range of standards (jointly with ISO/ITU-T) for video and audio compression. Although generic, the MPEG2 standard will be used mostly for applications requiring broadcast quality video, which can be achieved at about 6 Mbps.

MPEG3

Moving Picture Experts Group standard for video, 40 Mbps, aimed at high definition TV.

MPEG4

Moving Picture Experts Group formed in 1992 to look into digital video transmission at very low bit-rates (tens of Kbps). The resultant standard is intended to use fractal and wavelet transforms, instead of Discrete Cosine Transform (DCT).

MPLS

Multi-Protocol Label Switching (also known as Tag switching). A protocol based on BGP developed by Cisco that allows separation of traffic streams within a network of shared routers. For instance, with MPLS, IP packets are not forwarded based on IP addresses on a hop-by-hop basis (see *RIP*). Instead, they are forwarded along precomputed paths, generated on the routing information for a specific virtual network supported.

MPOA

Multiple Protocols Over ATM. A mechanism for providing LAN interconnectivity over an ATM network—see *LANE*.

MQSeries

An asynchronous message queuing product from IBM.

MRO

In the United States, the term "Maintenance, Repair, and Operations" (MRO) has come to be used to describe the major commodity products used by a large company. This might include anything from copier paper and paperclips through to higher-value items such as PCs.

MSAT

This was one of the first mobile satellite systems to deliver mobile communications directly from satellite to hand-held device. MSAT operates in the 1 to 2 GHz band.

MTA

Message transfer agent. The part of an *MTS* that gets the messages from one part of the network to another.

MTBF

Mean time between failures. A commonly used measure of the reliability of a computing or networking element. It gives a measure of how long to expect a particular component to work before keeling over. MTBF calculations should be featured as part of the design documentation for any network.

MTIE

Maximum Time Interval Error. Finds the peak-to-peak variations in the phase of a timing signal over a given observation period (synchronization performance measurement). Useful for specifying transients, bounding maximum wander, and controlling frequency offsets.

MTP

Message transfer part. The lower layers of the *C7*, which deal with link control and information transfer.

MTS

Message transfer system. One or more *MTAs* that combine to provide a store-and-forward messaging system between *UAs* and message switches.

MTTR

Mean time to repair. Like *MTBF*, taken as a measure of the reliability of a computer or networking element. The focus is more on availability this time, since MTTR tells the user how long to wait before normal service is resumed.

MTU

Maximum Transmission Unit. The largest possible unit of data that can be sent on a given physical medium. For example, the MTU of Ethernet is 1,500 bytes.

Mu-law

The North American standard for coding analog telephone calls onto a digital carrier. It uses non-uniform quantizing logarithmic compression and is equivalent to the similar but not compatible *A-law* used in Europe.

MULDEX

Multiplexer and demultiplexer.

Multicast

A special form of broadcast where copies of a packet are delivered only to a subset of all possible destinations. Hence, it describes the transmission of a message to a number of recipients (usually, a distribution list). A more discerning form of one-to-many distribution than a broadcast.

Multicast addressing

Ethernet addressing scheme used to send packets to devices of a certain type or for broadcasting to all nodes. The least significant bit of the most significant byte of a multicast address is 1.

Multilink PPP

Multilink PPP is a variant of PPP that addresses the additional features of compression and channel aggregation. PPP-ML, as it is known, is outlined in RFC 1717.

Multimedia

Human-computer interaction involving text, graphics, voice, and video. Often also includes concepts from hypertext. This term has come to be almost synonymous with CD-ROM in the personal computer world, because the large amounts of data involved are currently best supplied on CD-ROM.

Multiplexer

A device usually found within telephone networks that is used to combine a number of individual channels, such as combining a number of 64-Kbps channels on a 2-Mbps *bearer* (using time division multiplexing).

Multiplexers allow efficient use of transmission capacity and are increasingly used on private networks to share the cost of circuits across a number of applications.

Multiplexing

Combining several signals for transmission on some shared medium (e.g., a telephone wire). The signals are combined at the transmitter by a multiplexer (a "mux") and split up at the receiver by a demultiplexer. The communications channel may be shared between the independent signals in one of several different ways: time-division multiplexing, frequency-division multiplexing, or code-division multiplexing.

If the inputs take turns to use the output channel (time-division multiplexing), then the output bandwidth need be no greater than the maximum bandwidth of any input.

If many inputs may be active simultaneously, then the output bandwidth must be at least as great as the total bandwidth of all simultaneously active inputs. In this case, the multiplexer is also known as a concentrator.

Multiprocessing

Running multiple processes or tasks simultaneously. This is possible when a machine has more than one processor or processing is shared among a network of *uniprocessor* machines. See also *multitasking* and *multithreading*.

Multiprocessor

A single computer having more than one processor and capable of executing more than one program at once.

Multitasking

Also known as multiprocessing, multiprogramming, and concurrency. Multitasking performs (or seems to perform) more than one task at a time. Multitasking operating systems such as Windows, OS/2, or Unix give the illusion of running more than one program at once on a machine with a single processor. This is done by "time-slicing" or dividing a processor load into small chunks, which are allocated in turn to competing tasks.

The first multitasking operating systems were designed in the early 1960s. Under cooperative multitasking, the running task decides when to give up the CPU, and under preemptive multitasking, the scheduler suspends the currently running task after it has run for a fixed period (time slice) and starts or restarts another task.

Multithreading

A program execution environment that interleaves instructions from multiple independent execution "threads." This differs from multitasking in that threads (or "light-weight processes") typically share more of their environment with each other than do tasks under multitasking.

Threads may be distinguished only by the value of their program counters and stack pointers while sharing a single address space and set of global variables. This means that threads may be switched extremely quickly, because there is very little state to save and restore.

Multithreading can thus be used for very fine-grain concurrency at the level of a few instructions, and so can hide latency by keeping the processor busy after one thread issues a long-latency instruction on which subsequent instructions in that thread depend.

MVS

Multiple Virtual Storage. A widely used and long lived operating system for mainframe computers from IBM. MVS has developed over the years and is now known as OS390 in its latest form.

MXR

Mail exchange record. A *DNS* resource record type indicating which host can handle electronic mail for a particular domain.

Named Pipes

Part of Microsoft's *LAN Manager*—provides an interface for interprocessing communications and distributed applications. It is an alternative to *NetBIOS* and is designed to extend the interprocess interfaces of the operating system OS/2 across a network.

NAS

Network Application Support. The Digital Equipment Corporation strategy for unified software environment, including Applications Access Services, Communication Services, and Information and Resource Sharing Services.

NAT

Number Address Translation. The function of replacing a private IP address (as found in an intranet) with a registered one, before packets are forwarded to the internet. When remote servers respond, their IP packets are set to a registered destination address, so they can be correctly routed back through the internet to the intranet firewall (which usually carries out the address translation).

NCS

Netware Connect Services, software which public network providers can use to offer wide-area links between existing corporate LANs.

NCSA

National Center for Supercomputing Applications. The birthplace of the first version of the *Mosaic World Wide Web browser*, in Urbana, IL.

NDIS

Network Device Interface Specification, allows Ethernet cards to work with LAN operating systems such as Novell and Pathworks.

NDR

Network data representation. The "on-the-wire" data format used by *DCE*.

NDS

NetWare directory service. The directory service that forms part of Novell's *NetWare* system.

NET3

Norme Europenne de Telecommunications, the European-wide standard for ISDN. The trade name for the standard is EuroISDN.

NetBEUI

NetBIOS extended user interface. The network transport protocol used by all of Microsoft's network systems and IBM's *LAN* server-based systems. NetBEUI is often confused with *NetBIOS*. The latter is the applications programming interface and the former is the transport protocol.

NetBIOS

Network basic input/output system. An IBM-developed protocol that enables IBM *PCs* to interface with and have access to a network.

It consists of an *API* that activates network operations on IBM PC compatibles running under Microsoft's *DOS*. It is a set of network commands that the application program issues in order to transmit data to and receive data from another host on the network. The commands are interpreted by a network control program or network operating system that is NetBIOS-compatible.

Netiquette

An *Internet* term that has been coined to describe network etiquette. It covers the conventions of politeness recognized on *Usenet* and in mailing lists. The most important rule of netiquette is "Think before you post." If what you intend to post will not make a positive contribution to the newsgroup and be of interest to several readers, don't post it! There are whole books dedicated to Netiquette.

Netlett

The network dual of an applet. Instead of providing a machine independent application, a netlett is intended to be a network independent agent. It would typically be used to traverse a network and gather or deposit information on behalf of its creator.

Netmask

A 32-bit mask that shows how an *Internet address* is to be divided into network, subnet, and host parts. The netmask has 1s in the bit positions in the 32-bit address that are to be used for the network and subnet parts, and 0s for the host part. The mask should contain at least the standard network portion (as determined by the address's class), and the subnet field should be contiguous with the network portion.

Netscape Navigator

A *World Wide Web browser* from Netscape Communications Corporation. Despite the first versions being released (free to the Internet) only in late 1994, it is now the most popular of the net browsers.

Netscape evolved from NCSA Mosaic (with which it shares at least one author) and runs on the X Window System under various versions of Unix, on Microsoft Windows, and on the Apple Macintosh. It features integrated support for sending electronic mail and reading Usenet news, as well as *RSA* encryption to allow secure communications for commercial applications, such as exchanging credit card numbers with net retailers.

Netscape provides multiple simultaneous interruptible text and image loading, native in-line *JPEG* image display, display and interaction with documents as they load, and multiple independent windows. It was designed with 14.4-Kbps modem links in mind, since this is now commonly available to many home users.

NetView/6000

An IBM network management product that supports autodiscovery, fault and data polling and performance baselining, graphical display with multiuser view of topology, a network-wide inventory, and support for multi-protocols including SNMP, CMIP, XMP, SNA.

NetWare

A Novell *LAN* operating system and associated products. Novell is a major player in the world LAN server market.

Network

In general, a system of interrelated elements that are interconnected in a dedicated or switched linkage to provide local or remote communication (e.g., of voice, video, data) facilitating the exchange of information between end users with common interests.

Also, the set of switches, circuits, trunks, and software that make up a telecommunications facility. Examples of well-known networks that provide a particular service are the *PSTN* and the *PSDN*.

The *ISO* seven-layer model attempts to provide a way of partitioning a computer network into independent modules from the lowest (physical) layer to the highest (application) layer. Many different specifications exist at each of these layers.

Networks can also be classified according to their geographical extent: local-area, metropolitan-area, or wide-area network.

Network address

The network portion of an *IP* address. For a class A network, the network address is the first byte of the IP address. For a class B network, the network address is the first 2 bytes of the IP address. For a class C network, the network address is the first 3 bytes of the IP address. In each case, the remainder is the host address. In the *Internet*, assigned network addresses are globally unique.

Network interface

The circuitry that connects a node to the network, usually in the form of a card fitted into one of the expansion slots in the back of the machine. It works with the network software and operating system to transmit and receive messages on the network.

Network layer

The third lowest layer in the *ISO* seven-layer model. It determines the routing of data packets from sender to receiver via the data link layer and is used by the transport layer. An example protocol is *IP*.

Network management

A general term embracing all the functions and processes involved in managing a network, including configuration, fault diagnosis, and correction. It also concerns itself with statistics gathering on network usage. The five main functions of network management are:

- Accounting management—the process of identifying individual and group access to various resources either to ensure proper access capabilities (bandwidth and security) or to properly charge the various individuals and department for such access.
- Configuration management—the process of identifying, tracking, and modifying the setup of devices on the network. This category is extremely important for devices that come with numerous custom settings (e.g., routers and file servers).
- Fault management—the process of identifying and locating faults in the network. This could include discovering the existence of the problem, identifying the source, and possibly repairing (or at least isolating the rest of the network from) the problem.
- Performance management—the process of measuring the performance of various network components. This also includes taking measures to optimize the network for maximum system performance (periodically measuring the use of network resources).
- Security management—the process of controlling (granting, limiting, restricting, or denying) access to the network and resources thereon. This could include setting up and managing access lists in routers (creating firewalls to keep intruders out), creating and maintaining password access to critical network resources, identifying the points of entry used by intruders, and closing such entry points.

To support these functions, a range of standard minimal *MIBs* have been defined, and many hardware (and certain software, such as *DBMS*) providers have developed private MIBs (in *ASN.1* format), allowing them to be compiled for use in a network management system.

Network management systems use standard protocols (e.g., *SNMP*) to communicate with managed elements in their network. In theory, any SNMP manager can talk to any SNMP *agent* with a properly defined MIB.

Neural network

Usually used to mean an artificial or computer-based neural network. Biological neural networks are much more complicated in their elementary structures than the mathematical models and simple computing elements used for artificial neural networks.

An artificial neural network is an array of many very simple processors ("units" or "neurons"), each possibly having a small amount of local memory. Most neural networks have some sort of "training" rule whereby the weights of connections are adjusted on the basis of presented patterns. Neural networks are designed to learn from examples, just like we all learn to recognize cars from examples of cars.

Newsgroup

One of *Usenet's* huge collection of topic groups or forums. Usenet groups can be "unmoderated" (anyone can post) or "moderated" (submissions are automatically directed to a moderator, who edits or filters and then posts the results). Some newsgroups have parallel mailing lists for *Internet* people with no netnews access, with postings to the group automatically propagated to the list and vice versa. Some moderated groups (especially those that are actually gatewayed Internet mailing lists) are distributed as digests, with groups of postings periodically collected into a single large posting, complete with an index.

NEXT

Near-end crosstalk. This is the interference that arises between two or more transmitters sending signals into the same cable. It is usually the most significant noise source, which limits the reach of high-speed duplex transmission systems.

NFS

Network file system. A protocol developed by Sun Microsystems and defined in *RFC* 1094, which allows a computer to access files over a network as if they were on its local disks. This protocol has been incorporated in products by more than 200 companies, and is now a de facto standard. NFS is implemented using a connectionless protocol (*UDP*) in order to make it *stateless.*

NIC

Network Information Center. A source of a huge amount of information about the Internet and related networking issues.

NICAM

Near Instantaneously Companded Audio Multiplex. Standard used for high quality audio broadcasting.

NIS

Network information service. Another name for Sun Microsystems' Yellow Pages client/server protocol, which is used for distributing system configuration data such as user and host names between computers on a network.

Sun licenses NIS technology to virtually all Unix vendors. The original name, Yellow Pages, is a registered trademark in the United Kingdom of British Telecommunications plc for its commercial telephone directory, so Sun changed the name of its system to NIS. All of the commands and functions in NIS still start with "yp," though.

NIS+

An enhanced version of NIS. It has influenced the *XFN* standard.

N-ISDN

Narrowband ISDN. A network capable of offering a single narrowband interface to a user capable of carrying a multitude of services (e.g., high quality fax, telephony, data, telex).

NIST

National Institute of Standards and Technology. U.S. governmental body that provides assistance in developing standards. It was formerly the National Bureau of Standards.

NMF

Network Management Forum. A body with a remit considerably broader than its name suggests. The initial focus was on the interfaces and procedures needed to manage complex networks composed of many different elements. The fact that many of the components of modern networks are computers has focused the NMF on a service-based approach to the management of networks and systems covering the traditional concerns of both telecommunications networks and distributed computer systems.

There are over 100 members of the NMF and there is a balance between all of the major trading regions across the computing and telecommunications industry.

Since its inception, the forum has introduced the *CMIP/CMIS* standard for interfacing network elements and management systems, *SNMP/*CMIS Interworking, SPIRIT (which captures the service provider requirements for computing platforms), and Open Edge (a legacy system integration and migration technique).

NNTP

Network news transfer protocol. A protocol defined in RFC 977 for the distribution, inquiry, retrieval, and posting of Usenet news articles over the Internet. It is designed to be used between a news reader client and a news server.

NOC

Network operations center. A location from which the operation of a network or internet is monitored. Additionally, this center usually serves as a clearinghouse for connectivity problems and efforts to resolve those problems.

Node

Generic term used to refer to an entity that accesses a network name resolution.

Nonce

A process carried out between two computers (or, strictly speaking, processes running on two computers) that guarantees the authenticity of their communications.

Nondeterminism

A property of a computation that may have more than one result. One way to implement a nondeterministic algorithm is using backtracking; another is to explore all possible solutions in parallel.

Nonproprietary

Software and hardware that is not bound to one manufacturer's platform. Instead, this equipment is designed to an open specification so that it can accommodate other companies' products designed to that same specification.

The advantage of nonproprietary equipment is that a user has more freedom of choice and a larger scope. The disadvantage is that when it does not work, you may be on your own.

NOS

Network operating system. An operating system that includes software to communicate with other computers via a network (e.g., Berkeley System Distribution Unix, Novell *NetWare*, LANtastic, Microsoft *LAN Manager*).

Notification

An unsolicited message sent out by a process to inform interested parties that some event has occurred.

NP-complete

Nondeterministic polynomial time. A set or property of computational decision problems that is a subset of NP (i.e., can be solved by a nondeterministic Turing machine in polynomial time), with the additional property that it is also NP-hard. A solution for one NP-complete problem would solve all problems in NP. Many (but not all) naturally arising problems in class NP are in fact NP-complete.

NSF

National Science Foundation. A U.S. government agency whose purpose is to promote the advancement of science. The NSF funds science researchers, scientific projects, and infrastructure to improve the quality of scientific research.

NT

Windows New Technology, a 32-bit operating system from Microsoft. NT has been ported to Intel 86 machines, DEC Alpha, Sun SPARC, and others, but vendors need to recompile their applications to run on other NT platforms—not binary compatible, unless platform emulates DOS, as in DEC AXP PCs.

NT1

Network termination. A piece of equipment (usually on a user's premises) that provides network access via a standard interface.

More specifically, the NT1 is a part of the general structure of the *ISDN*. It terminates the line to the local exchange and presents a defined interface (known as the S/T reference point) to the user. If a user has ISDN-compatible equipment, it can be plugged straight into the NT1. If not, a *TA* is required between the terminal and the NT1.

NT2

Network Termination 2. A more complicated device than an NT1, typically found in digital PBXs, that performs layer-2 and layer-3 protocol functions as well as concentration services. An NT2 connects a TE1 or a TA to an NT1 and is part of the overall ISDN group of components and reference points.

NUA

Network Unit Address. If you are accessing an X.25 network then you will have a number (your NUA) identifying your equipment (e.g., your VDU or printer). Similar in function to a VTAM identifier.

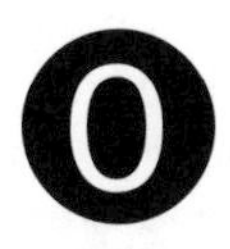

ABCDEFGHIJKLMNOPQRSTUVWXYZABCDEFGHIJKLMNOPQRSTUVWXYZABCDEFGHIJKLMNO

OA Office Automation. General term for mail and other facilities provided to speed commercial operations.

OASIS Open Automotive Services through Integrated Systems, an automatic invoicing system based on EDI*Net for fleet operator lease plans.

Object An *abstract*, encapsulated entity that provides a well-defined service via a well-defined interface. In the broadest sense, an object can be any thing, entity, concept, or abstraction (real or imagined) with clear boundaries and behavior. It is accepted that an object may be an instance of one (or more) classes of similar objects that share common attribute types and operations.

In *object-oriented* programming, an object is a unique instance of a data structure defined according to the template provided by its *class*. Each object has its own values for the variables belonging to its class and can respond to the messages (*methods*) defined by its class.

Object code The machine code generated by a source code language processor such as an *assembler* or *compiler*. A file of object code may be immediately executable or it may require linking with other object code files (e.g., libraries) to produce a complete executable program.

Object ID

Object identification. The name that uniquely distinguishes one object from all others. The short form of an object ID is unique within a local network. A longer form of the object ID may be required to uniquely identify it on a more widely distributed system. If the local machine name or *LAN* name is part of the object ID, special consideration is required to support object mobility. The universal naming of objects is one of the concerns tackled by *IDL* and *CORBA*.

Objectives

Goals based on concrete criteria (e.g., the objective of producing software that incurs no more than three major errors in the first year of operation).

Object model

A computer representation that encapsulates data attributes and behavioral processes (operations) for an *object*. Object model software may respond to events, triggers, and requests for service submitted as message stimuli (with a finite set of message types, argument types, and message formats). An object model is a graphical representation of the structure of objects in a system, including their identity, attributes, operations, and associations between objects.

Object-oriented

A philosophy that breaks a problem into a number of cooperating objects. Each object has defined properties (e.g., it can inherit features from another object). Object-oriented design is becoming increasingly popular for complex or distributed systems (e.g., in the development of network management tools).

Object program

The translated versions of a program that has been assembled or compiled.

Observational equivalence

Used to describe a situation in which two items are indistinguishable to an external observer. They may be very different internally, but the behavior they exhibit to the outside world is identical. Observational equivalence is useful in that it can be used to describe a desirable state of affairs in the design of system interfaces—that two items should match perfectly at the periphery. There is a mathematical definition of observational equivalence that allows useful theory to be derived for system analysis.

OC

Optical Carrier, the basis for the SONET transmission hierarchy:

OC-1 for 51.84-Mbps transmission systems, and at multiples of OC-1 for higher rate transmission, with some of the more common rates being:

OC-9 for 466.56 Mbps;

OC-12 for 622.12 Mbps;

OC-18 for 933.12 Mbps;

OC-24 for 1244.12 Mbps;

OC-36 for 1866.24 Mbps;

OC-45 for 2488.32 Mbps.

Occam

The language developed for use with the *transputer*. Occam was designed to exploit the component simplicity and inherent parallelism in transputer-based systems.

Occam's razor

A maxim, first proposed by the English philosopher William of Occam, that entities should not be multiplied more than necessary. That is, the fewer assumptions an explanation of a phenomenon depends on, the better it is. The sentiment is echoed in Einstein's words "systems should be as simple as possible but not simpler."

ODA

Open document architecture (formerly office document architecture). An *ISO* standard for describing documents. It allows text, graphics, and facsimile documents to be transferred between different systems in a common format. In practice, ODA is little used.

ODBC

Open Database Connectivity, an emerging standard database API from Microsoft, and the standard data access mechanism for Windows and Mac applications. The ODBC API is an alternative to SQL.

ODIF

Open document interchange format. Part of the *ODA* standard.

ODP

Open Distributed Processing. One of a number of organizations (most of which have the word "open" in their title) who provide standards and/or components that allow computers from different vendors to interwork. The work of ODP is focused on standardizing an *OSI* application layer communications architecture, thus allowing distributed systems to interwork.

ODP is a natural progression from *OSI*, broadening the target of standardization from the point of interconnection to the end system behavior. The end objective of ODP is to enable the construction of distributed systems in a multivendor environment through the provision of a general architectural framework to which such systems must conform.

One of the cornerstones of this framework is a model of multiple viewpoints, which enables different participants to observe a system from a suitable perspective and a suitable level of abstraction. In practical terms, ODP is one of a number of contributors (e.g., *ANSA*, *OMG*, *IAB*, *OSF*) in this area.

OEM

Original equipment manufacturer. A company that makes equipment (e.g., computers) as opposed to one that sells equipment made by other companies.

OFL

Percentage Overflow. The percentage of calls offered to a route and overflowed to a subsequent choice.

Oftel

Office of Telecommunications. The U.K. telecommunications regulator.

OLE

Object linking and embedding. A distributed-object system and protocol from Microsoft. OLE allows an editor to farm out some part of a document to another editor and then reimport it for inclusion in the final product. For example, a desktop publishing system might send some text to a word processor or a picture to a bitmap editor using OLE. See also the associated idea of *DDE*.

OLO

Other licensed operator, term applied to any telecommunications administration other than a national carrier (e.g., in the United Kingdom, any operator except BT—Energis or Mercury).

OLTP

Online transaction processing. The processing of transactions (usually very high volumes) by computers in real time. An instance would be a single-customer database dealing with inquiries from many remote terminals. Usually, an OLTP monitor would be used to interface the many users to the database.

OMG

Object Management Group. A consortium aimed at setting standards in object-oriented system design and programming. *CORBA* is one of the more widely known outputs from the OMG.

OMNI

Open management interoperability. One of the prime outputs from the *NMF.* OMNI comprises standards that represent the consensus view of a wide range of operators and manufacturers. The OMNI group is responsible for the OMNIpoint specifications, which include the *CMIS* and *CMIP* standards for connection between network elements and network management systems.

OMNIpoint1

A basket of specifications (standards, implementation specifications, testing methods and tools, and object libraries) approved by the Network Management Forum members. Represents a coherent set of functional specifications for network management but does not address the user interface, the MIB, or the APIs to be used with them. It does include definitions for the Common Management Services, Communications Services, and Managed Objects.

OMT

Object modeling technique. An application life cycle development method and graphical notation scheme that spans object models, dynamic models, and functional models from analysis through design and implementation. Basically, it is a framework within which to use objects.

OO

Object-oriented. The idea of computer analysis, design, and system development, where real-world concepts (like customers, orders, products) are modeled as encapsulated objects, each of which has specific attributes and operations. This approach contrasts with conventional computing systems design in that database design is usually isolated from program design.

Similar objects are grouped together in *classes* with common data attributes and operations that can be *inherited* by instances of the class. Objects communicate with other encapsulated objects by sending messages.

In theory, reusable subcomponent part objects can be assembled in various ways to define a wide variety of business object models, reducing reinvention of the wheel and incompatible applications. OO technology is now proving effective in the creation of innovative computer systems, communication networks, interface design, quality assurance, and (as a parallel development) reliable, reusable software modules.

OOP

Object-oriented programming. A programming approach in which an *object* is a data structure (abstract data type) encapsulated with a set of routines, called *methods*, which operate on the data. Operations on the data can only be performed via these methods, which are common to all objects that are instances of a particular *class* (see *inheritance*).

Thus, the interface to each object is well defined, which allows the code implementing the methods to be changed so long as the interface remains the same. Hence, there is a level of flexibility built into this way of doing things.

Each class is a separate module and has a position in a class hierarchy. Methods or code in one class can be passed down the hierarchy to a subclass or inherited from a superclass. This allows for rapid development and provides some level of protection against consistency errors.

Open Group

The organization that resulted from the amalgamation of the Open Software Foundation (OSF, the developers of the distributed computing environment, or DCE) and X-Open, the industry association that seeks to set standards for open computing (and owns the Unix trademark).

Open System

A general term for systems that are built with standard interfaces which allow components from different manufacturers to be connected together.

Operating system

Software such as *MVS* OS/2 (from IBM), *VMS* (from DEC), *DOS* (from Microsoft), and *Unix* (from many vendors) that manages the computer's hardware and software. Unless it intentionally hands control over to another program, an operating system runs programs and controls peripherals.

The operating system schedules tasks, allocates storage, handles the interface to peripheral hardware, and presents a default interface to the user when no application program is running. It may be split into a *kernel*, which is always present, and various system programs, which use facilities provided by the kernel to perform higher level housekeeping tasks, often acting as servers in a *client/server* relationship.

Some would include a graphical user interface and window system as part of the operating system, and others would not. The facilities an operating system provides and its general design philosophy exert an extremely strong influence on programming style and on the technical cultures that grow up around the machines on which it runs.

Over the last 30 years or so, there have been many operating systems: to the above list could be added AOS, CP/M, Chorus, GEORGE 3, Mach, Multics, Novell NetWare, OS/2, Pick, RISC OS, Solaris, VxWorks, Windows NT. Some of these are special-purpose, others are simply alternatives. Despite the wide variety on offer, only a handful are widely deployed.

Operational semantics

A set of rules specifying how the state of a computer changes while executing a program. The overall state is typically divided into a number of components (e.g., stack, heap, registers). Each rule specifies certain preconditions on the contents of some components and their new contents after the application of the rule. Other approaches to the same problem are axiomatic semantics and *denotational semantics*.

Operation code

The part of a machine code instruction that specifies the operation to be performed.

Oracle

One of the most widely used relational database management systems. The name is used to denote both the company and the product.

ORB

Object request broker. A piece of software that receives, redirects, and routes real-time interobject messages. An ORB allows developers to write objects without having to worry about specifics of the system they run on. The ORB worries about locating resources, thus allowing the system objects to be distributed anywhere.

OSCA

Open systems cabling architecture. A structured cabling system, primarily for local networks that use *UTP* cable, that is run from a central hub out to users. The hub provides a flexibility point and allows the installed star configuration to be reset as a bus or a ring as required.

OSCA

An architecture developed by Bellcore to promote the interoperability of large-scale software products (i.e., those that comprise many programs and transactions). The architecture is based on four main components—three layers and a fabric. The three layers are:

- The corporate data layer, which ensures the proper management of data. It separates data from processing and presentation.
- The processing layer, which provides number crunching, report generation, and the like.

- The user layer, which isolates software from varied user devices, thereby separating concerns and easing future evolution.

The fabric that holds these together is called "communications software," which provides the services and infrastructure for building the various layers.

The concepts introduced in OSCA have found their way into much of the modern thinking on how systems should be constructed.

OSF

Open Software Foundation. A foundation created by nine computer vendors (Apollo, DEC, Hewlett-Packard, IBM, Bull, Nixdorf, Philips, Siemens, and Hitachi) to promote open computing. The OSF aims to provide common operating systems and interfaces based on developments of *Unix* and the *X Window System* for a wide range of different hardware architectures. The OSF is the holder of the *DCE.*

OSI

Open Systems Interconnection. The ISO reference model for communicating systems, which consists of seven separate protocol layers. This model of network architecture, and the suite of protocols (protocol stack) to implement it, have been developed by ISO since 1978 as a framework for international standards in heterogeneous computer network architecture. The architecture is split between seven layers, from lowest to highest:

Layer 1, Physical layer;

Layer 2, Data link layer;

Layer 3, Network layer;

Layer 4, Transport layer;

Layer 5, Session layer;

Layer 6, Presentation layer;

Layer 7, Application layer.

Each layer uses the layer immediately below it and provides a service to the layer above. Within each of these layers, a number of protocols have been defined. The concept of the protocols is to provide manufacturers and suppliers of communications equipment with a standard that will provide reliable communications across a broad range of equipment types.

OSI is the umbrella name for a series of nonproprietary protocols and specifications, comprising, the OSI reference model, *ASN.1*, *BER*, *CMIP* and *CMIS*, *X.400* (*MHS*), *X.500* (directory service), Z39.50 (search and retrieval protocol used by *WAIS*), and many others.

OSPF

Open Shortest Path First. A routing protocol widely used in TCP/IP networks.

OSS

Operating support system. A general name for those systems needed to keep an organization going. It typically covers billing, network management, customer-handling, service management, ordering, and other key business functions.

Out-of-band signaling

The use of transmission facilities other than the primary channel bandwidth for the transmission of signaling and other control information. *ISDN* is a good example of out-of-band signaling—one of the basic philosophies in its design has been to separate the user and signaling plane.

Overload

A word that has many meanings. More technically, it means multiple definitions of an object operation. Different input arguments (signatures) requesting the same operation name (message type) cause different methods (functions) to be invoked.

It can also mean exceeding design limits, which is another way of saying that something is overengineered.

Oversubscription

In frame relay service definition, oversubscription occurs when the sum of the *CIRs* for all *PVCs* on a port exceeds the port connection speed. Subscription levels of 200% are typically allowed. Oversubscription is possible because of dynamic capacity allocation in modern data networks.

PABX (or PBX)

Private automatic branch exchange. A communication switch situated on a customer's premises that is used to connect customer telephones (and related equipment) to a local exchange. The PABX also provides cheap and fast internal calls within the customer's telephone system.

Modern PABXs offer numerous software-controlled features, such as call forwarding and call pickup. A PABX uses technology similar to that used by a main network switch, albeit on a smaller scale.

Packet

The unit of data sent across a network. Packet is a generic term used to describe a unit of data at any layer of the *OSI* protocol stack, but it is most correctly used to describe application layer data units. In practice, it is difficult to distinguish between a packet, a datagram, a cell, and a frame.

Packet Internet Groper

Abbreviated to "ping," which is a slang term for a short network message sent by a computer to check for the presence, connectedness, and alertness of another. The Unix command "ping" can be used to do this manually and to measure round-trip delays.

Packet switching

The mode of operation in a data communications network whereby messages to be transmitted are first transformed into a number of smaller, self-contained message units known as packets.

Packets are stored at intermediate network nodes (packet-switched exchanges) and are reassembled into a complete message at the destination. Each packet is routed to its destination through the most expedient route (as determined by some routing algorithm). This means that not all packets traveling between the same two hosts, even those from a single message, will necessarily follow the same route. So the destination computer has to reassemble the packets into their appropriate sequence. To do this, it has to buffer incoming packets (to account for the differing transit times) and ensure that each packet is correctly received.

Packet switching is used to optimize the use of the bandwidth available in a network. A widely used *ITU-T* recommendation standard for packet switching is *X.25.* Many countries have a well-established X.25 network (e.g., Transpac in France, PSS in the United Kingdom).

PAD

Packet assembly/disassembly facility. A device that converts a serial data stream into discrete packets in the transmit direction and converts the received packets back into a serial data stream. A PAD adds header information to the transmit packet to allow it to be routed to the proper destination. The *ITU-T* recommendations *X.28*, *X.29*, and *X.*3 defined PAD operations and interfaces.

PADs are used in the *X.25* packet-switched network, permitting terminals that cannot interface directly to the network to do so.

PAL

Phase Alternate Line, a German-originated technique for coding broadcast TV signals, a variant of NTSC. The television standard for most of Europe (excluding France which uses SECAM), Australia, New Zealand, and other countries. PAL displays 625 horizontal lines per frame, interlaced at 25 frames per second.

PAL+

Phase Alternation Line Plus. A widescreen TV broadcast picture format.

PAP

PPP Authentication Protocol, used to protect information delivered over dialup Internet connections.

Parallel processing

The simultaneous use of more than one computer to solve a problem. It is usually associated with computing-intensive tasks that can be split up into a large number of small chunks, which can be processed independently on an array of relatively inexpensive machines. Many engineering and scientific problems can be solved in this way. It is also frequently used in high-quality computer graphics.

There are many different kinds of parallel computer (or parallel processor). They are distinguished by the kind of interconnection between processors (known as processing elements) and between processors and memory.

Parallel computers are often classified by the way in which processors execute instructions—each processor executing the same instructions at the same time (single instruction/multiple data, or SIMD) or each processor executing different instructions (multiple instruction/multiple data, or MIMD).

Parameter

A variable whose value may change the operation but not the structure of some activity (e.g., an important parameter in the productivity of a program is the language used). Also commonly used to describe the inputs to and outputs from functions in programming languages. In this context they may also be known as "arguments."

Parse

The breaking of a language into its constituent elements. There are many computing applications that require a set of instructions to be understood by a program and this is often accomplished by parsing the sequence of instructions so as to make sense of them.

Pascal

Programming language named after the French mathematician Blaise Pascal, designed by Niklaus Wirth in the early 1970s. Pascal was designed for simplicity and for teaching programming as a reaction to the complexity of ALGOL 68. It emphasizes structured programming constructs, data structures, and strong typing.

Pascal has been extremely influential in programming language design and has a great number of variants and descendants.

Pathological

Use of a data set that is at (and beyond) the extremes of the typical. Pathological data are frequently used to test a system's response to exception conditions.

Pattern

An approach to solving a given type of problem by using analogy/comparison with established or existing solutions of a similar type (and application of templates or reference designs).

PBX

Private branch exchange. An alternative to (and frequently used diminutive of) *PABX*.

PC

Personal computer. Any computing system for use primarily by one person. The *IBM PC* is the most familiar and ubiquitous example of the PC.

PC DOS

Personal Computer Disc Operating System, IBM's version of MS DOS for the IBM PC.

PCI

A processor-independent PC bus from Intel. PCI is closely linked to the processor and requires that graphics systems, hard disk controller, and networking adapters are built into the motherboard. Intel offers royalty-free licenses on PCI, allowing manufacturers to produce PCI-compliant chip sets, boards, and peripherals.

PCM

Pulse-code modulation. A method by which an audio signal is represented as digital data. PCM is widely used within telephony networks.

It should not be confused with the common term for Sony's F1 format, which stores PCM digital audio on videotape.

PCMCIA

Personal Computer Memory Card International Association. Organization set up in 1989 to establish standards for personal computer cards—small plug-in units that provide additional memory and input/output functions, designed to be installed or removed without opening the computer case.

Three types of cards are currently specified. Type 1 is 3.3 mm thick and is designed to provide extra memory. Type 2 is 5 mm thick and is typically used for modems. Type 3 at 10.5 mm is intended for special needs such as high-capacity memory.

PCN

Personal communication network. A two-way digital cellular network based on *GSM*. It provides the capacity for the delivery of a wide range of mobile services. See also *PCS*.

P-code

An intermediate language developed in the 1970s that was used to enhance the portability of Pascal programs. The P-code provided a standard format for applications that could readily be used to generate the machine-specific object code on a variety of target computers. A modern reincarnation of the idea can be seen in the *Java* language.

PC Robot

A PC which is connected to a system as if it were an ordinary user terminal, but carries out sequences of actions automatically, often in the absence of a real human user.

PCS

Personal communication services. The range of services supported on the *PCN*. It includes mail, fax, and other data services, as well as voice. Not all PCS systems are GSM (e.g., those provided in the United States).

PCTE

Portable common tool environment. An *ECMA* standard framework for software tools developed in the *Esprit* program. It is based on object modeling and defines the way in which tools access and interwork with the environment.

PDA

Personal digital assistant. A small handheld computer used to write notes, track appointments, and otherwise organize affairs. PDAs provide all the functionality of a cheap pad of paper (at hundreds of times the cost currently) with more processing capability but far less storage capacity. All PDAs have data communications capacity (e.g., fax, e-mail), but no voice capability.

The best known of these devices is probably the U.S. Robotics PDA, which uses a combination of pen-based input and character recognition software to simulate the familiar pocketbook.

PDC

Personal digital cellular. One of a number of systems for cellphones. PDC originates from Japan and is similar in function to *DAMPS*, *TACS*, and *GSM*.

PDF

Portable Document Format. A PostScript-based file format from Adobe than can describe documents in a completely device- and resolution-independent manner. PDF is used in Acrobat and uses PostScript language to describe not only the visual aspects of a file, but also additional elements such as annotations, hypertext links, miniature "thumbnail" sketches, and bookmarks.

PDH

Plesiochronous digital hierarchy. A transmission system for voice communication using *plesiochronous* transmission. PDH is the conventional multiplexing technology for network transmission systems and is widely deployed by most of the public network operators. Service levels (i.e., multiplex speeds) within PDH are described with terms like *DS-1*.

Because of the difficulty of extracting data tributaries from the multiplex, PDH is being replaced by *SONET* and other *SDH* schemes.

PDL

Page description language. A general term for a specification for the interpretation of characters and their layout for printing.

PDU

Protocol data unit. A packet of data passed across a network. The term implies a specific layer of the *ISO* seven-layer model and a specific protocol.

Peer to peer

Communications between two devices on an equal footing, as opposed to host/terminal or master/slave. In peer-to-peer communications, both machines have and use processing power.

Perimeter network

A small, single-segment network between a firewall and the Internet for services that the organization wants to make publicly accessible to the Internet without exposing the network as a whole.

Perl

Practical Extraction and Report Language. A general-purpose language, often used for scanning text and printing formatted reports. It can be very concise (and thus hard to read) because of its powerful operators, such as regular expression substitution.

Persistence

(1) In programming parlance, a property of a programming language in which created objects and variables continue to exist and retain their values between runs of the program.

(2) The length of time a phosphor dot on the screen of a cathode ray tube will remain illuminated after it has been energized by the electron beam. Long-persistence phosphors reduce flicker, but tend to generate ghostlike images that linger on screen for a fraction of a second.

Petri net

A directed graph in which nodes are either places (represented by circles) or transitions (represented by rectangles). A net is marked by placing tokens on places. When all the places connected to a transition (the input places) have a token, the net is fired by removing a token from each input place and adding a token to each place pointed to by the transition (the output places). Petri nets are used to model concurrent systems, particularly operating systems and network protocols. Petri nets come in various flavors, including colored and multilevel.

PGP

Pretty Good Privacy. A high-security *RSA* public-key encryption application for MS-*DOS*, *Unix*, VAX/*VMS*, and other computers. It is distributed as freeware.

PGP allows people to exchange files or messages with privacy and authentication. Privacy and authentication are provided without managing the *keys* associated with conventional cryptographic software. No secure channels are needed to exchange keys between users, which makes PGP much easier to use. This is because PGP is based on public-key cryptography.

Phases

Individual stages of work on a piece of software (e.g., the testing phase). The idea of the software life cycle is that a development can be staged into a series of phases. Each phase can be managed as a separate entity. In practice, it is rarely that neat. Even so, structuring into phases gives some basis for defining quality gates between various activities.

Physical layer

The lowest layer in the ISO seven-layer model. It concerns electrical and mechanical connections and *MAC*, and is used by the data link layer. Example physical layer protocols are *CSMA/CD*, *token ring*, and bus.

Physical media

Any physical means for transferring signals between OSI systems. Considered outside the OSI model, and sometimes referred to as "Layer 0," or the bottom of the OSI reference model.

PICS

Protocol Implementation Conformance Statement. Formal means of identifying conformance to the protocol specifications.

PICS

Platform for Internet Content Selection. A voluntary mechanism for rating and assuring the quality/suitability of material placed on the Internet.

Ping

Packet Internet Groper. A program useful in testing and debugging *LAN/WAN* troubles. It sends out an echo and expects a specified host to respond in a specified time frame. The term is also used as a verb: "Ping host X to see if it is up." It was probably originally contrived to match the submariners' term for the sound of a returned sonar pulse.

Pink

Apple Computer's object-oriented operating system. It subsequently became known as *Taligent*, following a merger with IBM upon its development.

PIO

Programmed Input/Output. A method of data transfer in which the host microprocessor transfers data to and from memory via the computer's INPUT/OUTPUT ports. PIO enables very fast data transfer rates, especially in single-tasking operating systems like DOS.

Pipe

A method installed in most operating systems that is used by programs to communicate with each other. For instance, one of the main attributes of the OS/2 operating system is that pipes can be created quickly and easily.

When a program sends data to a pipe, the data are transmitted directly to the other program without ever being written into a file.

Pipeline

A sequence of functional units that performs a task in several steps like an assembly line in a factory. Each functional unit takes in inputs and produces outputs, which are stored in its output *buffer*, which is the next stage's input buffer. This arrangement allows all the stages to work in parallel, thus giving greater throughput than if each input had to pass through the whole pipeline before the next input could enter.

The costs of pipelining are greater *latency* and complexity due to the need to synchronize the stages in some way so that different inputs do not interfere. The pipeline will only work at full efficiency if it can be filled and emptied at the same rate that it can process. Pipelines may be synchronous or asynchronous. A synchronous pipeline has a master clock, and each stage must complete its work within one cycle. The minimum clock period is thus determined by the slowest stage. An asynchronous pipeline requires handshaking between stages so that a new output is not written to the interstage buffer before the previous one has been used.

Pixel

Picture element. The smallest resolvable rectangular area of an image, either on a screen or stored in memory. Each pixel in a monochrome image has its own brightness, from 0 for black to the maximum value (e.g., 255 for an 8-bit pixel) for white. In a color image, each pixel has its own brightness and color, usually represented as a triple of red, green, and blue intensities.

Pkunzip

A program written by PKWare and released on a freely distributable basis to encourage use of their professional compression system Pkzip. It is mainly for IBM PCs, but versions for other computers exist.

Pkzip

A file compression and archiver utility for MS-DOS and other computers from PKWare. It uses a variation on the sliding window compression algorithm. It comes with Pkunzip and Pklite and is available from most *FTP* archives.

Platform

The foundation of a system on which subcomponents depend. The term is often used in a vague way to describe a computer plus operating system plus some application software.

Plesiochronous

Nearly but not quite synchronous. Transmitted signals all have the same nominal rate, but is synchronized to different clocks, so there is inevitable drift between signals.

The core of many digital networks (include public telephone networks) are plesiochronous (see *PDH*). It is this arrangement that limits flexibility and fuels the drive for *SDH* and *SONET*.

Plug and play

Hardware or software that can be used (played) immediately after being installed (plugged in), as opposed to hardware or software that requires some degree of local configuration. Plug-and-play systems are the aim of open systems, in which components from a variety of suppliers should be readily interconnected to do what a user wants.

PM

Paid Minutes. A means of internetwork accounting. Rather than exchanging bills relating to every call, two international operators both keep a running total of the number of "paid minutes" of calls that they offer to the correspondent's network. Every month or so the two operators compare their totals to find the net transfer of paid minutes and settle up accordingly.

PM

Performance monitoring. Usually related to the measurement of network congestion and throughput.

PMR

Private mobile radio. Usually noncellular-based systems, standardized by *ETSI*, used for private data and voice exchange.

PNCA

Peer-to-peer network communication architecture. Objects that communicate in a network as equals, in contrast to a master/slave or client/server relationship.

PNG

Portable Network Graphic (pronounced "ping"). This is intended to be the replacement for the GIF format: it is technically more advanced and uses a public-domain compression algorithm.

The compression method used in PNG is similar to that used in Pkzip and related file-compression utilities. PNG uses a 7-pass interlacing scheme that sends the gross outline of the image first, followed by refinements that build up the final picture. It supports up to 48 bits per pixel and can store gamma information, ensuring, therefore, that the picture is correctly rendered on whatever type of machine is being used.

PNNI

Private Network-Node Interface. Allows multivendor switch interoperability for the setup of switched virtual circuits. Developed by the ATM Forum, it will eventually allow dynamic ATM networks to be constructed with heterogeneous (multi-vendor) components.

PNO

Public Network Operator.

PnP

Plug and Play. A standard, pioneered by Microsoft and endorsed by industry leaders. This standard hopes to address the problems of adding I/O adapters to a PC computer system. Adapters designed to the Plug and Play standard will self configure, and automatically resolve system resources such as interrupts (IRQ), DMA, port addresses, and BIOS addresses.

POCSAG

Post Office Coding Standards Advisory Group. An early (1982) standard for radiopaging that formed the basis for *ERMES*.

Pointer

An address from the point of view of a programming language. A pointer may be typed in order to indicate the type of object to which it points.

Point to point

Direct link between two points in a network or communications link.

Polling

Process of interrogating terminals in a multipoint network in a prearranged sequence by controlling the computer to determine whether the terminals are ready to transmit or receive. If one of them is ready, the polling sequence is temporarily interrupted while the terminal transmits or receives.

PON

Passive optical network. An arrangement whereby many users get their network connection via a single optical fiber. Splitters are used to get the required number of users onto a single fiber, which is then connected to a switch.

PON is seen by the providers of telephone services as a cheap provision option that also gives a high level of flexibility.

PoP

Point of presence. A site where there exists a collection of telecommunications equipment, usually modems, digital leased lines, and multiprotocol routers. The PoP is put in place by an *ISP*.

An ISP may operate several PoPs distributed throughout their area of operation to increase the chance that their subscribers will be able to reach one with a local telephone call. The alternative is for them to use virtual PoPs via some third party.

POP

Post office protocol. A *client/server* protocol that allows clients with dialup access to pick their mail up from a server. A POP provides an alternative for users without *SMTP* delivery. The latest version of POP (POP3) is not compatible with earlier versions.

Port

A device that acts as an input/output connection. Serial ports and parallel ports are both familiar examples.

The term also refers to the transport of software from one system to another system and the necessary changes that need to be made in order for the software to run correctly in its new environment.

Portfolio

A set of related systems. The management of a portfolio of software systems entails categorizing and prioritizing each one against overall business objectives.

Porting

Translating software to run on a different computer and/or operating system. Porting was once a major activity. The appearance of *APIs* has greatly reduced the pain of getting software that works on a range of computers.

In addition to a higher degree of consistency in platform interfaces, there are guides to the use of programming languages (such as XPG-3 from X/Open) that help to reduce the level of variability between software developers, hence easing the complexity of any porting that is required.

Port I/O address

A window through which software programs communicate commands to an installed host adapter board. The commands are communicated 8 bits at a time.

POSIX

Portable operating system interface for computer environments. A set of *IEEE* standards designed to ease the portability of applications between different operating systems. POSIX is contained in a number of standards: IEEE 1003.1 defines a Unix-like operating system interface, 1003.2 defines the shell and utilities, and 1003.4 defines the real-time extensions.

Postmaster

The electronic mail contact and maintenance person at a site connected to the Internet (or similar distributed mail network). Often, but not always, the postmaster is the same as the system administrator. The Internet standard for electronic mail (*RFC* 822) requires each machine to have a postmaster address.

PostScript

A *PDL* from Adobe Systems. PostScript is an interpreted language used as a PDL by the Apple LaserWriter, and now by many laser printers and on-screen graphics systems. Its primary application is to describe the appearance of text, graphical shapes, and sampled images on printed or displayed pages.

POTS

Plain old (or ordinary) telephone system. The basic services provided over the public switched telephone network.

PPP

Point-to-point protocol. Provides the Internet standard method for transmitting *IPU* packets over serial point-to-point links. It is defined in *RFC* 1171. As a means of accessing the Internet, PPP has a number of advantages over the alternative, *SLIP*. First, it is designed to operate both over asynchronous connections and bit-oriented synchronous systems. Second, it can configure connections to a remote network dynamically and test that the link is usable.

Presentation layer

The second highest layer (layer 6) in the *ISO* seven-layer model. Performs functions such as format conversion to try to smooth out differences between hosts. Allows incompatible processes in the application layer to communicate via the session layer.

PRI

Primary Rate Interface. A type of ISDN connection, as specified in the ITU I.431 standard. PRI consists of 30 B+D channels (in Europe, 2.048 Mbps) or 23 B+D channels (in North America and Japan, 1.544 Mbps). In both cases it runs over the same physical interface as the standard E1/T1 carrier.

PRI is typically used for connections such as one between a PBX and a network switch. It is also widely used to do videoconferencing.

Procedure

A method or set of steps defining an activity. Technically, a procedure is a program that can be executed as a subactivity by another program.

Procedure calls are described in term of message passing. A message names a method and may optionally include other arguments. When a message is sent to an object, the method is looked up in the object's class to find out how to perform that operation on the given object. If the method is not defined for the object's class, it is looked for in its superclass and so on up the class hierarchy until it is found or there is no higher superclass. Procedure calls always return a result object, which may be an error, as in the case in which no superclass defines the requested method.

Process

Technically, a procedure that is being executed on a specific set of data. More generally, a process is a procedure for doing something.

Processor

That part of a computer capable of executing instructions. More generally, it is any active agent capable of carrying out a set of instructions (e.g., a transaction processor for modifying a database).

Product

Usually, an entity to be sold. More generally, a product is the end result of some process.

Program

(1) A set of instructions for a computer arranged so that when executed they will cause some desired effect (such as the calculation of a quantity or the retrieval of a piece of data).

(2) Programs, data, designs for programs, specifications, and any of the other information that is relevant to a particular set of executable computer instructions (either existing or planned).

Programming language

An artificial language constructed in such a way that people and programmable machines can communicate with each other in a precise and intelligible way. FORTRAN, COBOL, and C are the three languages that account for most of the deployed software systems at present.

Project management

The systematic approach for analyzing, organizing, and completing a project.

Proprietary

Any piece of technology that is designed to work with only one manufacturer's equipment. The opposite of proprietary is open, the guiding principle behind Open Systems Interconnection.

Protocol

A set of formal rules describing how to transmit data, especially across a network. Low-level protocols define the electrical and physical standards to be observed, bit- and byte-ordering, and the transmission and error detection and correction of the bit stream.

High-level protocols deal with the data formatting, including the syntax of messages, the terminal-to-computer dialog, character sets, and sequencing of messages.

There are many different protocols for many different purposes. Most of those currently deployed are defined in the Internet *RFCs* or by standards bodies such as *ISO*.

Protocol converter

An application-specific node that connects otherwise incompatible networks. It converts data codes and transmission protocols to enable interoperability. It is also known as a *gateway*.

Protocol stack

A layered set of protocols that work together to provide a set of network functions. Each intermediate layer uses the layer below it to provide a service to the layer above. *ISO's* seven-layer model (Open Systems Interconnection) is an attempt to provide a standard framework within which to describe protocol stacks.

Prototype

A scaled-down version of something, built before the complete item is built, in order to assess the feasibility or utility of the full version.

Proxy ARP

The technique in which one host, usually a router, answers *ARP* requests intended for another machine. By faking its identity, the router accepts responsibility for routing packets to the real destination.

Proxy ARP allows a site to use a single IP address with two physical networks. In practice, this can be a complex (and risky) option—more systematic use of IP addresses usually works better.

Proxy gateway

A computer and associated software that will pass on a request for a *URL* from a *World Wide Web* browser such as *Netscape* to an outside server and return the results. This provides *clients* that are sealed off from the *Internet* with a trusted *agent* that can access the Internet on their behalf. The client's user should not be aware of the proxy *gateway*.

Browsers such as Mosaic and Netscape can be configured to use a different proxy or no proxy for each URL access method. So the access to *FTP*, *Go-*

pher, *WAIS*, news, and *HTTP* services can each be handled differently.

Proxy server

A process providing a *cache* of items available on other servers, which are presumably slower or more expensive to access.

PSDN

Public switched data network. The collection of interconnected systems that provides anyone served by the public network operator with a range of data services.

PSN

Packet-switched network. A network in which data are transmitted in units called packets. The packets can be routed individually over the best available network connection and reassembled to form a complete message at the destination.

In the United Kingdom, this ranges from low-speed packet access to *broadband* connection. It is fairly representative of the range of modern PSDNs, which use *X.25*, *frame relay*, *SMDS*, and *ATM*.

PSTN

Public switched telephone network. The collection of interconnected systems operated by the various telephone companies and administrations (*PTTs*) around the world. It is also known as *POTS* , in contrast to *ISDN*, which extends data as well as voice service to the end user.

The PSTN started as human-operated analog circuit-switching systems (manned by operators), progressed through electromechanical switches, and is now almost completely digital except for some final connections to the end user.

There are a number of things that limit the PSTN in terms of providing more of the exotic services promised by ISDN. These include code conversion (*A-law* to *mu-law*, or vice versa) on some international calls, robbed-bit signaling in North America (with 8 Kbps taken from a 64-Kbps link), data compression to save bandwidth on long-haul trunks, and signal processing such as echo suppression on some routes.

PTT

Post, telephone, and telegraph administration. The dominant or monopolistic public network operator (outside of the United States). PTT often implies state ownership and linkage with the postal service, although some are now privatized. PTTs provide the telecommunications services in a country (e.g., British Telecom in the United Kingdom and NTT in Japan).

Most PTTs outside the United States and United Kingdom are still government monopolies, but this situation is eroding. PTT is, to all intents and purposes, the same as a PNO (public network operator) or PTO (public telecommunications operator).

Public domain

The total absence of copyright protection. If something is in the public domain, then anyone can copy it or use it in any way. The author has none of the exclusive rights that apply to copyrighted work.

The phrase "public domain" is often used (incorrectly) to refer to freeware or shareware. The former is software that is copyrighted but is distributed without charge; the latter is similar but invites some later payment. Public domain spurns any sort of rights over whatever has been distributed.

Public-key encryption

An encryption scheme in which each person gets a pair of keys: the public *key* and a private key. Each person's public key is published, while the private key is kept secret.

Messages are encrypted using the intended recipient's public key and can only be decrypted using the private key. The need for sender and receiver to share secret information (keys) via some secure channel is eliminated: all communications involve only public keys, and no private key is ever transmitted or shared.

Public-key cryptography can be used for authentication (digital signatures) as well as for privacy (encryption). *RSA* is an example of a public-key cryptosystem.

PVC

Permanent virtual circuit. In data networking services, a circuit that is defined in a static manner with static parameters, but is not tied to a given physical path through the network. PVCs are often found as a part of large corporate networks, since it is convenient for the network provider to respond to customer demands without having to install fixed equipment. They can be provided via a range of technologies, *ATM* and *frame relay* being two of the more popular ones.

A B C D E F G H I J K L M N O P Q R S T U V W X Y Z A B C D E F G H I J K L M N O P Q R S T U V W X Y Z A B C D E F G H I J K L M N O

Q.1200 An ITU-T standard. General recommendations on telephone switching and signaling for Intelligent Networks.

Q.1208 An ITU standard. General Aspects of the Intelligent Network Application Protocol.

Q.1290 An ITU standard. Glossary of Terms Used in the Definition of Intelligent Networks.

Q.2761–2764 ITU standards for B-ISDN (originally B-ISUP).

Q.3 ITU standard for the TMN interface between a customer installation and service provider. Covers CMIS/CMIP and the Managed Objects between them. Q3 is the interface to the management network as distinct from the telecommunications network itself.

Q.700 ITU standard. Introduction to Signaling System No. 7. Consists of a general description of SS7 and SS7 networks.

QA Quality assurance. A formal approach to product development and delivery with the goal of zero defects. There are various standards by which organizations can be accredited (e.g., *ISO* 9001) as proof of their intent to adhere to QA principles.

Q adapter

In network management, the device that allows any element of the managed network to be connected to a network manager equipped with standard interfaces. The insertion of the adapter usually limits the range of actions that the management system can apply to the managed element.

QAM

(1) Quadrature amplitude modulation. A method for encoding digital data in an analog signal in which each combination of phase and amplitude represents one of sixteen 4-bit patterns. This is required for fax transmission at 9,600 bps. QAM is a bandwidth-efficient modulation scheme, in contrast to *QPSK*.

(2) Quality assurance management.

QPSK

Quadrature phase-shift keying. Like QAM, a method for encoding digital data for transmission over an analog link. QPSK does not have the bandwidth efficiency of QAM, but is more power-efficient.

Quality assessment

A systematic and independent examination to determine whether quality activities and related results comply with planned arrangements, and whether these arrangements are implemented effectively and are suitable for achieving objectives.

Quality of service

Measure of the perceived quality of a service. Usually based on tangible metrics such as time to fix a fault, average delay, loss percentages, and system reliability.

Quality surveillance

The continuous monitoring and verification of the status of procedures, methods, conditions, processes, products, and services and the analysis of records in relation to stated references to ensure that specified requirements for quality are being met.

Quality system

The organizational structure, responsibilities, procedures, processes, and resources for implementing quality management.

Quality system standard

A document specifying the elements of a quality system. The *ISO 9001* standard (which is generally used to control software development) is a widely known and used quality standard.

Queue

A facility that stores transactions or event-oriented messages and activates them for processing in a specific sequence. There are different ways of implementing queues, such as FIFO (first in, first out), priority, and event type. The term *buffer* is usually synonymous with queue.

Quicktime

Apple Computer's standard for integrating full-motion video and digitized sound into application programs.

QWERTY

The standard English-language typewriter keyboard as opposed to the Dvorak or foreign-language layouts. QWERTY is the layout of the first six keys on the top row of letters.

Historical note: The origins of the QWERTY layout is much misunderstood. It is sometimes said that it was designed to slow down the typist, but this is wrong. In fact, it was designed to speed up typing by ensuring that the mechanical levers of the typewriter did not jam. This involved laying the keyboard out so that the levers carrying the letters did not interfere with each other in normal use. The jamming problem has long since passed with the advent of the computer, but the keyboard layout lives on.

ABCDEFGHIJKLMNOPQRSTUVWXYZABCDEFGHIJKLMNOPQRSTUVWXYZABCDEFGHIJKLMNO

R

The ISDN Reference Point that lies between a TE2 (a piece of non-ISDN equipment) and a Terminal Adaptor.

RACF

Remote Access Control Facility. A large system security product from IBM. Originally it was intended for MVS only but has subsequently been ported to run on other operating systems.

RAD

Rapid application development. A loose term for any software life cycle designed to shorten development times. RAD is associated with a wide range of approaches to software development including prototyping with interface builders, *CASE* tools, iterative *life cycles*, workshops, constrained time development, and reuse of applications, templates, and code. RAD usually involves the customer as the key part of the decision-making process and a facilitator to guide the development process.

RAID

Redundant array of inexpensive disks. There are a variety of RAID options (from RAID 0 to RAID 5), all of which describe ways of configuring disk drives in a way that improves the availability of data.

Simple duplication of disks, also known as mirroring, is included here, as is the idea of *striping*, where data are spread across a number of disks. In addition to speeding up access (especially for read operations), striping gives the option of adding a check disk that can be used to recreate data in the event that one of the other disks fails. So, for instance, a disk could be added at the end of a line of five disks.

For every stripe of data put on the five storage disks, a parity check is written to the sixth. In the event of failure, the lost data can then be recovered. This arrangement gives as secure an arrangement as more traditional mirroring with, in this case, six disk drives instead of ten.

RARP

Reverse address resolution protocol. A protocol defined in *RFC* 903 that provides the reverse function of *ARP*. RARP maps a hardware address (*MAC* address) to an Internet address. It is used primarily by diskless nodes when they first initialize to find their Internet address.

Raster

The area of a video display that is covered by sweeping the electron beam of the display in a series of horizontal lines from top to bottom. The beam then returns to the top during the vertical flyback interval.

Rate adaptation

A means of carrying data of a lower rate than the channel they are carried over. For instance, *ISDN* channels running at 64 Kbps can readily connect two terminals running at 19.2 Kbps as long as they agree on a rate adaptation scheme. The two common standards for this are V110 and V120, both from the *ITU*. The former is more prevalent in Europe, the latter in the United States.

RBOC

Regional Bell operating company. There were seven Baby Bell companies created by the 1982 *MFJ*, which specified the terms of the AT&T divestiture. These were NYNEX, Bell Atlantic, Bell South, Southwestern Bell, U S West, Pacific Telesis, and Ameritech. The first two have subsequently merged.

RDBMS

Relational Database Management System. A structured computer information source and retrieval systems that conforms to the principles first put forward by Codd and Date, where the basic unit is a table with rows and columns. Data are defined, accessed, and modified with SQL statements.

Real time

Not a very precise term, but generally taken to mean the rapid transmission and processing of event-oriented data and transactions as they occur. It contrasts with data processing systems, where information is being stored and retransmitted or processed as *batches*.

Real-time systems are required for monitor and control systems, but are not required when long response times (e.g., overnight) are acceptable.

RED

Random Early Detect. In an IP network, RED provides the ability to flexibly specify traffic handling policies so that throughput under congested conditions is maximized.

Redundancy

(1) The provision of multiple interchangeable components to perform a single function in order to cope with failures and errors. Redundancy normally applies primarily to hardware. For example, one might install two or even three computers to do the same job. There are several ways these could be used. They could all be active all the time, thus giving extra performance through parallel processing as well as extra availability; one could be active and the others simply monitoring its activity so as to be ready to take over if it failed (termed "warm standby"); or the spares could be kept turned off and only switched on when needed (cold standby). Another common form of hardware redundancy is disk mirroring.

Redundancy can also be used to detect and recover from errors, either in hardware or software. A well-known example of this is the cyclic redundancy check, which adds redundant data to a block in order to detect corruption during storage or transmission. If the cost of errors is high enough (e.g., in a safety-critical system), redundancy may be used in both hardware and software with three separate computers programmed by three separate teams and some systems to check that they all produce the same answer or some kind of majority voting system.

(2) In communications, redundancy is the proportion of a message's gross information content that can be eliminated without losing essential information. It is carried to protect against corruption—redundant bits in a data packet can be used to detect and sometimes correct transmission errors.

Regression

Used to denote repetition during the testing of *software* systems. Regression is an important aspect of locating the errors introduced by change.

Relation

A two-dimensional table with rows and columns in an *RDBMS*. A table must have at least two columns. Each row represents one relationship between column values. An example would be a table that relates customer names to their telephone numbers.

Relational database

A database based on the relational model developed by E. F. Codd. A relational database allows the definition of data structures, storage and retrieval operations, and integrity constraints.

In a relational database, the data and relations between them are organized into tables. A table is a collection of records and each record in a table contains the same fields. Certain fields may be designated as keys, which means that searches for specific values of that field will use indexing to speed them up. Records in different tables may be linked if they have the same value in one particular field in each table.

INGRES, Oracle, IDMS, and Microsoft Access are well-known examples of relational databases. Commercial offerings of this type are very mature and can hold huge volumes (terabytes) of data.

Relationship

The concept, central to object orientation, that describes any type of association existing between two or more objects (e.g., inheritance, requires, provides, uses). All relationships are bidirectional (e.g., is_son_of and is_father_of).

Remote echo

A copy of the data being received is returned to the sending system for display on the screen.

Remote login
Operating interactively on a remote computer using a protocol over a computer network as though locally attached. Remote login is usually done via programs such as rlogin or *Telnet.*

Remote networking
Extending the logical boundaries of a corporate LAN over wide-area links to give remote offices, teleworkers, and mobile users access to critical information and resources.

Removability
A feature where the media in a removable media disk drive can be removed, then replaced with the same or different media without causing problems to the operating system. If removability was not supported, media in a removable media drive could not be removed without potential loss for data unless the computer was turned off.

Rendering
The preparation of an image for display on a chosen display device. This includes handling characteristics of that device that do not match those of the image (e.g., dithering a 24 bit image onto an 8 bit palette).

Repeater
Device that connects digital network cable segments. Regeneration and retiming ensure that the signal is clearly transmitted through all segments. The functionality is defined in detail in the IEEE 802.3 specification.

Repository
A data store holding (or pointing to) software and systems entities that designers and developers could reuse in the process of delivering new "systems solutions." The repository provides services to manage the creation, use, versions, maintenance, translation, and viewing of these entities.

Repudiate
Denial by a message originator that a message has been sent, or by a receiver that it was received.

Requirements analysis
The analysis of a user's needs and the conversion of them into a statement of requirements prior to specification. There are many systematic methods that can be used to carry out structured requirements gathering.

Reseller

A long-distance carrier that does not own a network, but leases bulk capacity and resells portions of it at a higher rate. It is also known as rebiller.

Resolve

Translates an *Internet* name into its equivalent *IP* address or other *DNS* information.

Reuse

The process of creating software systems using existing artifacts rather than starting completely from scratch. Code, components, designs, architectures, operating systems, and patterns are all examples of artifacts that can be reused. Also methods and techniques to enhance the reusability of software.

There are several sub-categories of reuse, the main ones being: generative reuse—reuse of the process (e.g., method, design, tools) used to create a component; local reuse—reuse of components within a product or product line or by a small team in several products (people working closely together will have implicit knowledge of how to use the component); domain reuse—systematic reuse of well understood common components across a specific area of interest, often in specific environments (people working in the domain will have implicit knowledge of how to use the component), sometimes called vertical reuse; global reuse—widespread reuse across domains, organizations, environments, and geography. All the knowledge needed to use the component has to be made explicit. Sometimes called horizontal reuse.

Reverse engineering

The process of analyzing an existing system to identify its components and their interrelationships and create representations of the system in another form or at a higher level of abstraction. It is usually undertaken in order to redesign the system for better maintainability or to produce a copy of a system without access to its source.

REXX

Restructured Extended Executive. A scripting language from IBM.

RFC

Request for comments. One of a series begun in 1969 of numbered *Internet* informational documents and standards widely followed by commercial software and freeware in the Internet and *Unix* communities. Early RFCs were discussion documents and ideas, but more recently they have become the source of implementation detail that supplements more general guidance of ISO and other formal standards.

Only a few of the RFCs are used as standards, but all Internet standards are recorded in RFCs. Perhaps the single most influential RFC has been RFC 822, the Internet electronic mail format standard. Some of the others worthy of notes are:

RFC 1014	External Data Representation (*XDR*);
RFC 1058	Routing Information Protocol (*RIP*) (updated by RFC 1388);
RFC 1081	*POP3*, Post Office Protocol version 3;
RFC 1094	Sun Microsystems's Network File System (*NFS*);
RFC 1112	*MBONE*;
RFC 1119	Network Time Protocol;
RFC 1156	*MIB* I—Management Information Base;
RFC 1157	Simple Network Management Protocol;
RFC 1171	Point-to-Point Protocol (*PPP*);
RFC 1213	MIB II—Management Information Base;
RFC 1388	An update to RFC 1058, the RFC defining routing information protocol;
RFC 1432	*Gopher;*
RFC 1441	*SNMP* v2;
RFC 1452	The RFC describing coexistence between SNMP v1 and SNMP v2;
RFC 1475	TP/IX protocol;
RFC 1521	Multipurpose Internet Mail Extensions (*MIME*);
RFC 1630	Universal Resource Identifiers;

RFC 792	Internet Control Message Protocol (*ICMP*);
RFC 903	Reverse Address Resolution Protocol (*RARP*);
RFC 959	File Transfer Protocol (*FTP*);
RFC 1155-7	SMI, Structure of Management Information (subset of ASN.1);
RFC 1334	PPP Authentication Protocols;
RFC 1661	The Point-to-Point Protocol (PPP);
RFC 1717	The PPP Multilink Protocol (MP);
RFC 768	Unacknowledged Datagram Protocol;
RFC 791	Internet Protocol (IP);
RFC 822	Internet e-mail address standard;
RFC1034	Domain Names—Concepts and Facilities;
RFC1035	Domain Names—Implementation and Specification.

RFCs are unusual standards in that they are floated by technical experts acting on their own initiative and reviewed by the Internet at large, rather than formally promulgated through an institution such as ANSI. For this reason, they remain known as RFCs even once adopted as working standards.

The RFC tradition of pragmatic, experience-driven, after-the-fact standards writing done by individuals or small working groups has important advantages over the more formal, committee-driven process typical of ANSI or ISO.

RFC

Request for Change. Documents raised by users, developers, and internal customers to identify minor product changes or improvements to be addressed in a future build of a product.

RFI

Request For Information. The first part of a tendering exercise, the RFI seeks ideas from prospective suppliers.

Ring latency
The time required for a signal to propagate once around a ring in a token ring or IEEE 802.5 network.

Ring topology
Network topology in which a series of repeaters are connected to one another by unidirectional transmission links to form a single closed loop. Each station on the network connects to the network at a repeater.

RIP
Routing information protocol. A distance vector (as opposed to link state) routing protocol. RIP is an *Internet* standard interior gateway protocol defined in RFC 1058 and updated by RFC 1388. RIP is based on *TCP/IP* and provides a link-state protocol that supports the exchange of information between hosts and gateways. It uses message broadcasts to find the optimum route to a destination based on a hop count.

RIPE
Reseaux IP Européens. An organization that allocates *IP* addresses. RIPE and the *InterNIC* are the bodies that administer IP addresses.

RISC
Reduced instruction set computer. A processor whose design is based on the rapid execution of a sequence of simple instructions rather than on the provision of a large variety of complex instructions.

RJ45
Connection standard based on U.S. telephone connector, commonly used for unshielded twisted pair cables used on a LAN.

rlogin
Remote login. *BSD* Unix utility to allow a user to log in on another host via the network, rather like Telnet. rlogin works by communicating with a daemon on the remote host.

RMON
Remote monitoring management information base. An extension of the *SNMP MIB* II, which provides a standards-based method for tracking, storing, and analyzing remote network management information. It was developed by the *IETF*.

ROM
An acronym for Read Only Memory. This is generally a chip on a computer or I/O card with software

programmed inside of it that controls some function or functions.

Root bridge Appointed by the spanning tree and used to determine which managed bridges to block in the spanning tree topology.

ROSE Remote operations service element. An *ISO* protocol that forms part of the applications layer. It provides facilities for initiating and controlling operations remotely. ROSE is a sublayer of layer 6 (presentation layer) in the OSI seven-layer model, providing *SASE* for remote operations.

Route The sequence of hosts, *routers*, *bridges*, *gateways*, and other devices that network traffic takes from its source to its destination. It is also a possible path from a given host to another host or destination.

Router A device that forwards packet-based traffic across networks. Routers operate at level 3 of the *OSI* model. They tend to be protocol-specific and act on routing information carried by the communications protocol in the network layer. A router is able to use the information it has obtained about the network topology and can choose the best route for packets to follow. Routers are independent of the physical level (layer 1) and can be used to link a number of different network types. This contrasts with a *bridge*, which is used to link segments of the same type. See also *brouter*.

Routing The selection of a communications path for the transmission of information from source to destination.

Routing bridge A MAC-layer bridge that uses network layer methods to determine a network's topology.

Routing protocol Protocol that accomplishes routing through the implementation of a specific routing algorithm.

Routing table A table stored in a router or some other internetworking device that keeps track of routes (and, in some cases, metrics associated with those routes) to particular network destinations.

Routing update

Message sent from a router to indicate network reachability and associated cost information. Routing updates are typically sent at regular intervals and after a change in network topology.

RPC

Remote procedure call. A protocol that allows a program running on one host to cause code to be executed on another host without the programmer needing to explicitly code for this. RPC is an easy and popular option for implementing the *client/server* model of distributed computing.

An RPC is implemented by sending a request message to a remote system (the server) to execute a designated procedure using parameters supplied, and a result message is returned to the caller (the client). There are many variations and subtleties in various implementations, resulting in a variety of different (usually incompatible) RPC protocols.

RPP

Relative Processor Power, a measure of processing capability used on MVS machines.

RS-232

The most common serial line standard. RS-232 is the *EIA* equivalent of *ITU-T* V.21. It uses 25-way D-type connectors but often only three wires are connected—one to ground (pin 7) and one for data in each direction. A computer (termed a *DCE*) RS-232 interface should have a female connector and should transmit on pin 2 and receive on pin 3. A terminal (termed a *DTE*) should have a male connector and should transmit on pin 3 and receive on pin 2.

The rest of the pins in the specified connector are related to hardware handshaking between sender and receiver and to carrier detection on modems. RS-423 specifies the electrical signals to go with the RS-232 connector.

RS-449

The serial line connection standard, seen as the successor to RS-232, as it eases the physical limitations, specifically the cable distance limitation. RS-449 has 37 pins, a 4,000-feet range (maximum with twisted pair) and 10 Mbps maximum transmission rate.

RSA
A well-known and used software-based public-key encryption method. It is named after its inventors—Rivest, Shamir, and Adleman.

RTF
Rich text format. An interchange format from Microsoft for exchange of documents between Word and other document preparation systems.

RTP
Real Time Protocol. An Internet transport protocol for real time traffic (such as Internet phone calls) that allows separate media streams to be handled individually.

Rules driven
A data driven system where the logic in the data is expressed using syntactic constructs. Rules driven applications are built in the same way as software interpreters.

Run time
The period of time during which a program is being executed, as opposed to compile time or load time.

Run-time system
The complete set of software that must be in primary storage while a user program is being executed.

ABCDEFGHIJKLMNOPQRSTUVWXYZABCDEFGHIJKLMNOPQRSTUVWXYZABCDEFGHIJKLMNO

S — ISDN reference point between TE1 (user terminals) and NT2.

SAA — System application architecture. A set of parameters for programmers to ensure that their software is IBM system–compatible.

SACSE — Signaling Association Service Control Element, an ISO standard.

SAP — Service access point. In the *OSI* layered model of communications, a boundary point where services from one layer are addressed to another.

SAP — Service advertisement protocol. Network servers running Novell's NetWare regularly broadcast information to every other device on the network that advertises the service that they have on offer.

SAP — Speech application platform. A general-purpose system component that is capable of providing *IVR* over a telephony network. An SAP guides callers by prompting them to enter dialed digits in response to the options they hear. It is an increasingly common part of *CTI* systems.

SAPI — Service Access Point Identifier. Part of the reference model used for signaling standards, the SAPI tells you where information is exchanged between neighboring protocol levels.

SASE

Specific application service element. A collection of interface definitions that form part of the *application layer* (the seventh and highest layer in the OSI model). It supports specific services such as *file* and job transfers that allow operations to be performed on remote files as if they were local. *FTAM*, *MHS*, and *CMIS* are all covered by SASE.

SATAN

Security administration tool for analyzing networks. Probes networked systems to see if services such as *FTP* are correctly set up, if well-known security flaws are present, and where potential attacks may come from. The results of a SATAN analysis are stored in a *database* for subsequent viewing with a standard *HTML browser*. A script called "repent" is included with the system to revise its name to SANTA.

SCAI

Switch/computer applications interface. Like *CSTA*, a standard that allows developers to produce *CTI* applications that reside on a network-connected switch. SCAI differs from CSTA in that it has been developed through ANSI and is intended for both public and private network operators.

Both SCAI and CSTA are aimed at network owners and operators rather than end users. To date, neither has been fully implemented.

Scalability

The ability to add power and capability to an existing system without significant expense or overhead. An "economy of scale" exists when a small increase in load produces a less-than-linear increase in overhead. A "diseconomy of scale" exists when a small increase causes a significant increase in overhead. With systems becoming so diverse, complex, and interrelated, scalability is a major issue in their design.

SCCP

Signaling connection and control point. A low-level part of the interexchange common-channel signaling system number 7 (*C7*).

SCCS

Source code configuration system. A facility that is available on the *Unix* operating system for keeping control of the *versions* and *variants* of *code* modules as they are developed. SCCS is one of the more basic (but also widely used) packages for *configuration management* during software development.

SCE

Service creation environment. Graphically based user software for entering complex enhanced service specifications.

SCP

Service control point. This is one of the central ideas behind intelligent networks. An SCP is one component (in practice, a powerful computer or set of computers) that responds to user service requests as they pass into the network. The SCP enables the network to appear intelligent as it can (1) act on the format, content, code or protocol of transmitted information, (2) provide additional or restructured information, and (3) use subscriber information with stored data (i.e., network information) by, for instance, translating dialled freephone numbers into ordinary telephone numbers.

In terms of its position within an intelligent network, the SCP is connected to service switching points (SSP), which in their turn are connected to the switch fabric of the network itself.

Screen scraping

A method of accessing a server in which the *client* presents itself as being a direct interface to a human user. The client "reads" information from the "screen" presented by the server and "sends" information as "keystrokes" from the pretend user.

Scripting

The use of a programming language to automate functions. There are a number of special purpose scripting languages (such as perl and Tcl) that allow the flow of network controls and computer operations to be directed as desired.

SCSI

Small computer system interface. A bus-independent standard for system-level interfacing between a computer and an intelligent device (e.g., an external disk). It is pronounced "scuzzy."

SCSI-1

Allows asynchronous transfers at up to 1.5 MBps and synchronous transfers at up to 5.0 MBps. Asynchronous is a classic Request/Acknowledgment handshake.

SCSI-2

With SCSI-2, asynchronous transfers can run at up to 3.0 Bps and synchronous transfers at up to 10.0 MBps. Synchronous is based on a Request/Acknowledgment scheme, except that it allows you to issue multiple requests before receiving the matching acknowledgments. In practice this means that synchronous transfers are approximately three times faster than asynchronous.

SCSI device

Device such as a host adapter board, fixed disk drive, or CD-ROM drive that conforms to the SCSI interface standard and is attached to a SCSI bus cable. The device may be an initiator, a target, or capable of both types of operation.

SCSI overhead

This is the time it takes for the host adapter to internally process a SCSI command.

SDH

Synchronous digital hierarchy. An international digital telecommunications network hierarchy that standardizes transmission around the bit rate of 51.84 Mbps, which is also called STS-1. Multiples of this bit rate comprise higher bit rate streams—so STS-3 is 3 times STS-1, STS-12 is 12 times STS-1, and so on. STS-3 is the lowest bit rate expected to carry *ATM* traffic, and is also referred to as STM-1 (synchronous transport module—level 1).

SDH specifies how payload data are framed and transported synchronously across *fiber-optic* transmission links without requiring all the links and nodes to have the same synchronized clock for *data transmission* and recovery (i.e., both the clock frequency and phase are allowed to have variations, or be *plesiochronous*). SDH offers several advantages over the current multiplexing technology, which is known as plesiochronous digital hierarchy. Where *PDH* lacks built-in facilities for automatic management and routing and locks users into proprietary methods, SDH can improve network reliability and performance, offers much greater flexibility and lower operating and maintenance costs, and provides for a faster provision of new services.

Under SDH, incoming traffic is synchronized and enhanced with network management bits before being multiplexed into the STM-1 fixed-rate frame. The fundamental clock frequency around which the SDH or *SONET* framing is done is 8 kHz or 125 μs. *SONET* is the American version of SDH.

SDK

Software developer's kit. The set of tools included in a software product that allows its integration, configuration, and customization. The usefulness of the product is often dependent on the quality of the associated SDK.

SDL

Specification and Description Language. One of the most enduring notations for the design of *state*-based software. SDL is particularly suitable for describing telecommunications systems, where the concept of signals causing system components to react in some way is fundamental. There are many SDL support tools (e.g., for analysis, code generation). SDL is defined as an international standard by the *ITU*.

SDLC

Serial data link control. A widely used protocol that operates at layer 2 in the *OSI* seven-layer model. That is, it is designed to implement the reliable transfer of *packets* from one point in a network to the next. SDLC was designed with facilities for error and *sequence checking* and is part of *SNA*. *HDLC* is the nonproprietary equivalent of SDLC.

SECAM

Sequential Couleur a Memoire. The broadcast television standard used throughout France.

SECAMS

Satellite Equipment Control And Monitoring Systems.

Second-generation language

Assembly language, the first move to a computer programming language oriented toward humans rather than computers. The *assembler* allowed the programmer to reason in terms of values, registers, and locations instead of low-level bits and bytes. Because of its speed and efficiency, many systems still contain significant portions of assembly code.

Second-party assessment
Assessments of contractors/suppliers undertaken on behalf of a purchasing organization. This may include assessment of companies or divisions supplying goods or services to others within the same group.

Secure logging
A method whereby an audit trail of system activity is received from a *bastion host* and placed in a secure location.

Security
Control mechanisms that prevent unauthorized use of resources. As well as the customary physical protection, most computer-based systems use some form of logical protections against misuse—passwords, coding, or *encryption*. It is also usual to have various levels of privilege (e.g., basic access, *system administration*). The general rule throughout is that users should have something (e.g., a *smart card*) and know something (an *access code*) before they are allowed to get hold of information.

SEE
A set of management and technical tools to support software development, usually integrated in a coherent framework. SEE is equivalent to an *IPSE*.

Segment
An electrically contiguous piece of the network in a bus topology *LAN*. Individual segments can be interconnected with repeaters, bridges, or routers.

Segmentation
In network terms, this is the splitting of an overloaded ring into two or more separate rings, linked by a bridge/router or multipurpose hub.

Segmentation
The process of dividing a program into sections (segments modules), which may be independently executed or changed.

SEI/CMM
A capability assessment and rating scheme developed at the Software Engineering Institute at Carnegie Mellon University. Its aim is to assess the capability of a software organization (from 1 = initial to 5 = optimal). At present, over 75% of software suppliers reside within level 1 of the defined levels.

In many ways, SEI/CMM (and its derivatives—such as *SPICE*) complements *ISO 9001* in that it provides a measure of how effective a supplier is, in contrast with the ISO focus on how efficient they are.

Semantics

The meaning of a language. When applied to programming languages, semantics is the meaning of a string (i.e., what the code will do) as opposed to the syntax (which states how symbols can be used legally). Semantics comes in two flavors—*denotational* and *operational.* In both, it tends to be couched in mathematical terms and is usually the preserve of the specialist.

Semaphore

The classic method of restricting access to shared resources (e.g., a *database*) in a multiprocessing environment. It was invented by E. Dijkstra and was first used in an early *operating system* (called T.H.E.). A semaphore is usually a protected variable (with only a few states) to which access is limited.

Sendmail

A very widely used part of the *Internet's electronic mail* system. Sendmail was built for the routing of mail over *TCP/IP* connections using *SMTP.* It is correctly termed a "mail transport agent" and will work in conjunction with other delivery agents that do not use SMTP. It is normally invoked via a user interface such as a mail program. Sendmail was developed as part of one of the early Unix implementations at *Berkeley.*

Sequencing

The process of dividing a *message* into packets for *transmission* so that reassembly can take place in the correct order. Each packet, block, or frame is given a sequence number so that the complete message can be put back at the receiving end. A variety of protocols take care of filling in any lost packets.

Serial interface

An interface which requires serial transmission, or the transfer of information in which the bits composing a character are sent sequentially. Implies only a single transmission channel.

Serializing

Another term for *marshaling.*

Serial transmission

Every bit of information follows another, rather than having simultaneous transmission of a number of bits. Serial transmission is the norm for data communications, parallel for the exchange of information between computers and peripherals.

Server

An object that is participating in an interaction with another object (usually a client) and is taking the role of providing the required service. It is one half of a *client/server* system.

Server clustering

The placing of all the servers on one or more rings in a central location.

Server farm

A cluster of servers in a centralized location serving a wide user population. A server farm would be used, for instance, to host a variety of pages served over the World Wide Web.

Server type

The class of a server within a client/server architecture. The main server types are name, directory, authentication, access control, cryptographic, communications, date, time, file, data, print, mail, *EDI*, applications, and presentation. There is no correspondence between server type and physical machines. It is possible that all of the above could run on one computer.

Service

An independently useful and well-defined function. Service is a common term in telecommunications, where service management and service creation are important commercial concerns.

Service creation

The provision of a new service for the user of a network. Typical services provided by network providers would be toll-free calls, charge card calls, messaging services, and the like. Service creation involves the reconfiguration of the software systems that control the network—some of these are in the network switch, some in associated system such as an *SCP*.

Service management

This is the care and maintenance of user network services. Service management is a major concern for agencies providing telecommunications services to third parties (i.e., *facilities management* companies and *telcos*).

Session

Generally, the connection of two *nodes* on a network for the exchange of data—any live link between any two data devices. In *SNA*, session is a basic concept that refers to a logical network connection between two addressable units (e.g., a mainframe and a terminal) for the exchange of data.

Session layer

The establishment, management, and termination of connections between different application programs in the *OSI* layered model of communication. The session layer assumes that a network connection is already in place.

SET

Secure Electronic Transfer. A dual signature system for money transactions over the Internet that allows each party to see only what needs to be seen. Jointly developed by Microsoft and Visa.

SGCP

Simple Gateway Control Protocol. End-to-end connections for voice in the packet network are established using this mechanism to set up connections in IP networks. SGCP was devised by Bellcore and is a UDP-based transaction protocol that permits manipulation of the connections represented by network endpoints (physical or logical). The connections are described using attributes such as IP addresses. SGCP manages call setup requests and connections from phones connected to the residential gateways.

SGML

Standard Generalized Markup Language. An international standard encoding scheme for linked textual information. *HTML* is a subset.

Shared tenant service

This is the provision of *PABX* services (frequently by a landlord) to multiple customers located in the same building, campus, or group of buildings. External calls can be placed and received over common lines and intracompany calls can be made without the use of outside local exchange carrier lines. It is most prevalent in the United States, where regulations frequently restrict the provision to protect the interests of the local carrier.

Shareware

Software that can be loaded from an open source, such as the Internet. It is usual for some license to be payable on such software (often no more than a few dollars or a crate of wine), which distinguishes it from *freeware*. See also *public domain*.

Shell Script

An interpreted UNIX program written in a high level scripting language such as perl.

.sig file

Or more correctly, .signature file. A file that automatically appends users' details (e.g., name, affiliation, motto) to their electronic communications.

Signal

Any event (e.g., a tone, frequency shift, binary value, alarm, message) that causes a change in *state* within a system.

Signaling

The passing of information and instructions from one point to another for the setting up or supervision of a telephone call or message transmission. There are many different types of signaling, depending on the required resilience, reliability, and breadth of information to be carried. Among the more prominent signaling systems are CCITT *SS7* (between telephone exchanges) and *remote procedure calls* (for computer-to-computer communications).

SIM

Subscriber identification module. A card that identifies users to their *GSM* phone. The card carries personal and bill data. The phone only works when the card is inserted.

SIMM

Single In-line Memory Module. A typical SIMM consists of a number of DRAM chips on a small Printed Circuit Board, or PCB, which fits into a SIMM socket on a computer's system board.

Simplex

One-way transmission path, with no response of any kind possible, as opposed to *duplex*, where a two-way dialog is established.

S-interface

One of the reference points in the *ISDN*. It is the standard end-user interface to basic rate ISDN, between the terminal and the *NT2*.

SIP

SMDS Interface Protocol, between the customer's equipment and the switching system in the telecommunications carriers' network.

SITA

Societe Internationale de Telecommunication Aeronautique. The worldwide airline reservation and ticketing service. It is probably the most extensive enterprise network in the world.

Site certification

This is a technique by which a trusted third party uses cryptographic keys to issue certificates to both the server site and the client in such a way that the client is assured of the servers authenticity. The need for site certification arises from the concerns when using the World Wide Web—to ensure that the site you are accessing is truly what it says it is (especially when valuables are being transferred).

SLA

Service level agreement. An agreement between a customer and a supplier that defines the range of services to be provided, both in scope and performance standard. A good SLA covers not only what should happen when things go well, but also when and how to escalate issues when there are problems.

Sliding window

Some (basic protocols) wait for an acknowledgment that one piece of information has been received before sending the next. A *protocol* that uses a sliding window increases efficiency by allowing a number of pieces of information to be in transit at one time. The maximum number in transit is known as the window size.

SLIP

Serial Line Internet Protocol. Software allowing the *IP*, normally used on *Ethernet*, to be used over a serial line, such as an *RS-232* serial port connected to a *modem*. It is defined in RFC 1055. SLIP modifies a standard *internet datagram* by appending a special SLIP END character to it, which allows datagrams to be distinguished as separate. SLIP requires a port configuration of 8 data bits, no parity, and *EIA* or hardware *flow control*. SLIP does not provide error detection, being reliant on other high-layer protocols for this. Over a particularly error-prone dialup link, therefore, SLIP on its own would not be satisfactory. A SLIP connection needs to have its IP address configuration set each time before it is established, whereas *PPP* can determine it automatically once it has started.

Slotting

The process of assigning a *circuit* to available *channel* capacity across a network (usually during the network design process).

SLU

Secondary logical unit. An *SNA* concept. The *LU* that contains the secondary half session for a particular LU–LU session. This is rather like client/server operation in that a given LU can be secondary for one session and primary for another.

Smart card

A credit card size device that has an onboard processor as opposed to the magnetic stripe on many conventional plastic cards. It can therefore be used for a variety of functions such as user authorization and cash transactions. Smart cards are widely used in France (where they were developed in 1981 in conjunction with Philips).

SMDS

Switched multimegabit data service. A *broadband* communications standard for the public network that does not require predefinition of a specific path (i.e., it is a *connectionless* service).

Smileys

A popular way of conveying emotion in an online session. Smileys originated in the Internet community, but have been more recently and broadly accepted as *emoticons.* A typical smiley is :-) (tilt head left to view). There are scores of these, from the quizzical to the miserable.

SMP

Symmetric multiprocessing. The use of a balanced set of computing elements, usually to give high-performance processing.

SMS

Server Systems Management, formerly Hermes. A Microsoft application that allows software distribution, PC software auditing, remote operation of PC (e.g., for problem diagnosis). Essentially a specialized application running on a server.

SMS

Short Message Service, for GSM and PCN. Allows messages of up to 160 characters to be sent from a PC to a pager. SMS is an embedded facility in commercial mail packages, rather like the established facility for sending faxes.

SMTP

Simple mail transfer protocol. The *Internet* standard for the transfer of mail messages from one processor to another. The protocol details the format and control of messages.

SNA

Systems Network Architecture. An *IBM* layered communications protocol, format, logical structure, and operational sequencing for sending information between disparate hardware and software. SNA defines three layers of communications—application, function management, and transmission subsystem—which allow communications concerns to be separated and users to be unaffected by the specifics of how the communicating parties talk.

SNAP

Sub Network Access Protocol. An Internet protocol that operates between a network entity in a subnetwork and a network entity in the end system. SNAP specifies a standard method of encapsulating IP datagrams and ARP messages on IEEE networks.

SNMP

Simple network management protocol. A *network management* tool used to manage customer network equipment and processes. It usually takes the form of a graphic on an *X Window* display. Formally, SNMP consists of three parts: structure of management information (SMI), *MIB*, and the protocol itself. The SMI and MIB define and store the set of managed entities, and SNMP transports information to and from these entities.

In operation, SNMP uses three basic request primitives: set, get, and get-next for configuration and performance information, and one asynchronous notification: trap for alarm and status information.

Originally designed to work with *TCP/IP*-based networks, SNMP does much the same job as the *CMIP/CMIS* standard. SunNet Manager, HP

OpenView, and IBM NetView/6000 are the most popular software products.

SNP

Subnetwork Protocol. A protocol that resides in the subnetwork layer below IP to provide data transfer through a local subnet. In some systems, an adapter module must be inserted between IP and the Subnetwork Protocol to reconcile their dissimilar interfaces.

Socket

The *Berkeley* Unix mechanism for creating a virtual connection between processes. Sockets form the interface between Unix standard input/output and network communication facilities such as TCP/IP.

At the simplest level, an application opens the socket, specifies the required service, binds the socket to its destination, and then sends or receives data.

Sockets can be of two types, stream (bidirectional) or *datagram* (fixed-length destination-addressed messages). The socket library function socket creates a communications end point or socket and returns a file descriptor with which to access that socket. The socket has associated with it a socket address, consisting of a port number and the local host's network address.

SODIMM

Small Outline DIMMs. Another type of memory commonly used in both notebook and laptop computers. It is like a 72-pin SIMM in a reduced-size package, but with some important technical differences. It is the way the pins are arranged that differentiates these two types of memory. The SO-DIMM supports 32-bit transfers and was designed for use in notebook computers.

Software

All programs (plus documentation) that are associated with a computer or computer-based system, as opposed to hardware, which is the physical equipment.

Software engineering

The development and use of systematic strategies (which are themselves often software-based) for the production of good-quality software within budget and on time.

Software glue

Common term for the software that binds together applications, operating systems, and other elements. The closest practical instantiation of software glue is probably the middleware developed for distributed systems (e.g., DCE, CORBA) under the auspices of the OMG and others.

Software package

A fully documented program (i.e., with user and *maintenance* manuals as well as systems design documentation). Alternatively, a set of programs designed to perform a particular task.

Software quality

A formal approach to software development, automated regression testing, configuration management, versioning, profiling, and release control with the goal of zero defects. *ISO 9001*, *TickIT,* and *SEI/CMM* all define procedures and give guidelines for software quality.

SOM

IBM's object-oriented development environment, which allows users to put together class libraries and programs. Associated with SOM is an *OMG CORBA*-conformant object request broker (known as *DSOM*) for building distributed applications. See also *COM*, the Microsoft equivalent of SOM.

SONET

Synchronous optical network. A standard that was developed by the Exchange Carriers Standards Association for optical-fiber transmission on the public network. It is intended to be able to add and drop lower bit-rate signals from the higher bit-rate signal without needing demultiplexing. SONET defines a set of transmission rates (e.g., *OC*-1), signals, and interfaces for fiber-optic transmission from 52 Mbps to 13.22 Gbps.

Source program

A program as written by the programmer using a programming language (e.g., *Ada*, *FORTRAN*, *C*, or *COBOL*). It must be assembled, *compiled*, or *interpreted* into *object* code before it can be executed on a computer. Many use the source program as the documentation for the software it describes.

Spamming

The distribution of unwelcome or unsolicited information (usually advertising) over *Internet* mail or newsgroups. Spamming is a gross breach of *netiquette* and usually attracts retribution.

Spanning tree

A technique that detects loops in a network and logically blocks the redundant paths, ensuring that only one route exists between any two LANs. Used in an IEEE 802.1d bridged network.

SPEC

Standard Performance Evaluation Corporation. An IT industry consortium which develops standard benchmarks, used to test hardware and operating system performance rather than database or throughput of transaction processing systems.

Specification

A description of a system or program that states what should be provided, but does not necessarily provide information on exactly how the system or program will work.

SPECmarks

A measure of processing power. The number of SPECmarks awarded to a processor give an indication of its power.

SPICE

Software process improvement and capability evaluation. SPICE is a development of ISO 9001 and a number of company-specific schemes. It aims to produce a set of standards and guidance documents to assist procurers, developers, suppliers, and maintainers in improving the quality and maintainability of systems and software on time and within budget. SPICE is being developed as an ISO standard.

SPID

Service Profile ID, uniquely identifies a B channel on the ISDN network. The SPID must be stored in any device accessing the ISDN.

SPIRIT

Service Providers Integrated Requirements for Information Technology. An initiative that came from the *NMF*. It is a consortium of *telcos* aided by their major *IT* suppliers, in which they jointly specify a general-purpose computing platform. Rather than producing standards, SPIRIT's aim is to provide guidelines that help a buyer select appropriate system components.

Split horizon

A routing technique where information about routes is denied to exiting router interfaces through which that information was received. Useful in preventing routing loops.

Spoofing

The use of a forged IP source address to circumvent a firewall. The packet appears to have come from inside the protected network, and to be eligible for forwarding into the network.

Spooler

Simultaneous peripheral operation online. The ability to store information in a local memory buffer for subsequent use. It is typically used to send files to print.

SQL

Structured query language. A widely used means of accessing the information held in a *relational database*. SQL enables a user to build reports on the data held in a computer via standard commands (e.g., insert, select, update, delete). There are many sophisticated tools based on SQL that allow *screens* of information to be linked to database entries.

SS7

CCITT signaling system number 7. A standard for interexchange signaling (also known as *C7*) that speeds up call processing by operating *out of band*, as opposed to being carried in the same stream as the traffic. SS7 includes many features, such as fraud detection, caller identification, store and forward, ring back, and concurrent data.

SSL

Secure sockets layer. A protocol for the secure transmission of information over the *Internet*. SSL was devised by Netscape Communications, which built the leading *World Wide Web browser*, to enhance network security.

SSP

Service switching point. The part of an intelligent network that intercepts and diverts requests that have to be serviced by an *SCP*. The SSP provides the routing function and ensures that decision making is passed to the *SCP* for action.

STA

Spanning Tree Algorithm. A function of managed bridges which allows redundant bridges to be used for network resilience, without the broadcast storms associated with looping. If a bridge fails, a new path to a redundant bridge is opened.

Stack

A group of network devices that are logically integrated into a single system.

Star topology

Network configuration where all the nodes are connected to a central point via individual cables.

State

Formally defined as the current condition(s) or value(s) stored in the data attribute(s) of an object. More casually, the state of a program or of a network defines what has happened so far and what can happen next. For instance, "waiting for digits" is a fairly straightforward state that a network switch might find itself in. The concept of state is very useful in the design of many systems, since it allows behavior to be analyzed in terms of stimulus (i.e., signal or message) and response (change of state). There is a significant amount of useful theory behind state machines, standard notations (*SDL*), and a number of support tools.

Stateful

When applied to a server, implies that the server maintains knowledge about and context for its clients between requests.

Stateless

When applied to a server, implies that a server maintains no knowledge about its clients between requests. Each request is handled in isolation from all preceding and following requests.

Statistical multiplexer

Usually called a "statmux." This device combines one physical channel with several data channels taking advantage of the fact that they each have different traffic characteristics. In effect, the statmux optimizes the use of communications capacity.

STB

Set-top box, a device that connect a conventional television to a data network so that it can display online and interactive services. Consortia such as DAVIC have established standards for set-top boxes, and products that deliver Internet access via a TV and Video on Demand are now available.

STM

Synchronous Transport Module. The link layer for the synchronous digital hierarchy, SDH. The speeds associated with the various STM levels are:

- STM level 1, bit rate 155.52 Mbps. Same as SONET STS-3c.
- STM-4, bit rate 622.08 Mbps. Same as SONET STS-12c.
- STM-8, bit rate 1.24416 Gbps.
- STM-12, bit rate 1.86624 Gbps.
- STM-16, bit rate 2.5 Gbps (2.48832). Same as SONET STS-48c

Store and forward

A technique used in data communications in which messages or packets are stored at an intermediate node in a network and then forwarded to the next *routing* point, when an appropriate line becomes free.

Stovepipe

A system—often old, dedicated, and proprietary—that operates outside of the cooperating mechanisms around which other systems are built. In order to interwork with stovepipes, integration techniques such as *screen scraping* are necessary.

Stream

(1) An *abstraction* referring to any flow of data from a source (or sender, producer) to a single sink (or receiver, consumer). A stream usually flows through a channel of some kind, as opposed to packets which may be addressed and routed independently, possibly to multiple recipients. Streams usually require some mechanism for establishing a channel or a connection between the sender and receiver.

(2) In the *C* language's buffered input/output library functions, a stream is associated with a file or device that has been opened using fopen. Characters may be read from (written to) a stream without knowing their actual source (destination), and buffering is provided transparently by the library routines.

(3) One of the leading computer suppliers, Sun Microsystems, has called its modular device driver mechanism STREAMS, leading to some confusion.

(4) In IBM's *AIX* operating system, a stream is a full-duplex processing and data transfer path between a driver in kernel space and a process in user space.

Striping

In *RAID*, the spreading of a data item across a number of disk drives. As well as allowing for consistency checks and removing reliance on one disk, striping evens out access demands over a set of drives and can speed up access (particularly when reading data).

STS

Synchronous Transport Signal. The logical signal specification for the SONET frame structure. This specifies electrical transmission rates based on 51.84 Mbps (STS-1). The other SONET transmission rates are:

- STS-3 at 155.52 Mbps (3 x STS-1).
- STS-3c at 155.52 Mbps (concatenated, not 3 x STS-1). Same as SDH STM-1.
- STS-12 at 622.08 Mbps (4 x STS-3).
- STS-12c at 622.08 Mbps (concatenated, not 4 x STS-3). Same as SDH STM-4.
- STS-48 at 2488.32 Mbps (4 x STS-12).
- STS-48c at 2488.32 Mbps (concatenated, not 4 x STS-12). Same as SDH STM-16.

Stubs

Stubs are the pieces of code on either side of a distributed application that provide the illusion of a remote operation happening as a local procedure call. There are two distinct sorts of stub code, one on the client side and the other on the server. The client stub provides a set of locally callable procedures whose signatures match those in the interface definition. When a remote procedure call is made, the client stub communicates with the server stub, which arranges for the real implementations of those procedures—supplied by the application developer—to be invoked on the server. Stub code is generated from interface definitions by a tool usually known as a stub or *IDL* compiler.

Subnet

An extension of the *IP* address that allows a network to be an autonomous entity and still be a part of a larger user network. It is typically used by large organizations to establish an internal network with often selective connection to the rest of the Internet.

Subscriber

An end user, customer, or local user.

SuperJANET

A high-speed successor to the U.K. *JANET*. It is based on *SMDS* and provides a platform for the transmission of complex multimedia data between researchers.

Support

Usually refers to an organization that has been set up to help with any problems found in an operational system. Support is usually structured into first line (a telephone help line to provide temporary fixes and advice), second line (who can fix configuration faults), and third line (who can fix problems in the core of the system).

SVC

Switched virtual circuit. The usual type of circuit for a packet-switched (*X.25* or *frame relay*) data connection. The user perception is of a dedicated physical link, but it is really a set of shared resources and the information built into the packets that provide the connection.

SWIFT

Society for Worldwide Interbank Financial Telecommunications. The authority over the system used for global electronic money settlements.

Switch

A device that responds to originator signals and dynamically connects callers to their desired destination. Generally used by telecommunication network designers to denote an *exchange*, such as Nortel DMS-250, GPT System X, and Ericsson AXE-10. All of these are capable of interconnecting thousands of users. The term is also used for smaller scale voice elements such as *PBXs* and data elements such as *routers*.

Switched Ethernet

An Ethernet hub with integrated MAC layer bridging or switching capability to provide each port with 10 Mbps of bandwidth. Separate transmissions can occur on each port of the switching hub, and the switch filters traffic based on destination MAC address.

Switched network

A network that is shared by several users, any of whom can establish communication with any other by means of suitable interconnections (switching) operations.

Switched resellers

Resellers that use their own switching hardware (and sometimes their own lines) and the lines of other interconnection services to provide long-distance service to their subscribers. They provide their own billing and customer services. The term only applies in the United States at present.

Switched services

All dialup long-distance services including conventional residential and wide-area transport services (most have incremental use charges).

Switched Virtual LAN

A logical network consisting of several different LAN emulation domains controlled through an intelligent network management application.

Switching

Process by which transmissions between terminals are interconnected. Switching is usually effected at nodal points in the network (switches or exchanges).

Switching hubs

Hubs that use intelligent Ethernet switching technology to interconnect multiple Ethernet LANs and higher-speed LANs such as FDDI.

Switchless reseller

A reseller of long-distance services that does not use any of its own lines, or switching equipment. All actual service and equipment is handled by the interconnection providers. Customer billing is usually done by the reseller itself. As with *switch reseller*, the terms applies only in the United States.

Switch site

A location that supports the establishment of a dynamic communication path. Usually realized as an exchange or a router.

Synchronization

The actions of maintaining the correct timing sequences for the operation of a system.

Synchronous

A form of communication transmission with a direct timing relationship between input and output signals. The transmitter and receiver are in sync and signals are sent at a fixed rate. Information is sent in multibyte packets. It is faster than *asynchronous* character transmission, since start and stop bits are not required. It is used for mainframe-to-mainframe and faster workstation transmission.

Synchronous transfer

A method of SCSI data transfer. With this type of data transfer, the SCSI host adapter and the SCSI device agree to a transfer rate that both support (this is known as synchronous negotiation). With this type of data transfer method, transfer rates of 5 MBps or 10 MBps (for Fast SCSI) are common.

Synchronous transmission

Transmission between terminals in which data are normally transmitted in blocks of binary digit streams and transmitter and receiver clocks are maintained in synchronism.

Syntax

The set of rules for combining the elements of a language (e.g., words) into permitted constructions (e.g., phrases and sentences). The set of rules does not define meaning, nor does it depend on the use made of the final construction.

System

(1) A collection of independently useful *objects* that happen to have been developed at the same time.

(2) A collection of elements that work together, forming a coherent whole (e.g., a computer system consisting of processors, printers, and disks).

System design

The process of establishing the overall *architecture*, blueprint, or layout of a system.

System design

The process of establishing the overall (logical and physical) architecture of a system and then showing how real components are used to realize it.

System integration

The process of bringing together all of the components that form a system with the aim of showing that the assembly of parts operates as expected. This usually includes the construction of components to carry out missing or forgotten functions and glue to interconnect all of the components.

System integrator

A vendor that offers design, connection, implementation, and management services for diverse network resources.

ABCDEFGHIJKLMNOPQRSTUVWXYZABCDEFGHIJKLMNOPQRSTUVWXYZABCDEFGHIJKLMNO

T.120 ITU-T videophone standard that covers data sharing/transfer, remote operation, whiteboard, text-talk, and so on.

T.30 ITU standard for Group 3 Facsimile that describes how machines communicate; Annex C is a proposal for adapting the Group 3 protocol to run on ISDN or digital private circuits (i.e., full duplex 64-Kbps channel).

T1.413 ANSI standard for ADSL that describes various channelization schemes.

T1 An AT&T term for a digital carrier facility used to transmit a *DS-1* formatted digital signal at 1.544 Mbps, widely used as part of the U.S. *PDH.* It is capable of carrying traffic equivalent to 24 multiplexed voice-grade channels. T1 transmission uses an *AMI* line coding scheme to keep the dc carrier component from saturating the line.

T2 The equivalent of four multiplexed *T1* channels.

T3 A trunk circuit used to transmit a *DS-3* formatted digital signal at 44.736 Mbps. Generally used with T1 circuits to carry traffic across the U.S. public switched telephone network. Capable of carrying the equivalent of 28 multiplexed T1 channels.

T4

A very-high-speed trunk circuit running at 274 Mbps, the equivalent of six multiplexed T3 channels.

TA

Terminal adapter. Equipment used to adapt *ISDN BRI* channels to existing terminal equipment standards such as *RS-232* and V.35. A TA is typically packaged like a modem, either as a stand-alone unit or as an interface card that plugs into a computer or other communications equipment (e.g., a router or *PBX*). A TA does not interoperate with a modem —it replaces it.

TAC

Terminal access controller. A device that connects terminals to the Internet, usually using dialup modem connections.

TACS

Total Access Communications System. A derivative of the AT&T-developed analog cellular radio standard (*AMPS*) adopted by U.K. Cellnet and Racal-Vodaphone, operating at 900 MHz and based on time-division multiple access.

Tagged queuing

A SCSI-2 feature that increases performance on SCSI disk drives. With tagged queuing, the host adapter, the host adapter driver, and the hard disk drive work together to increase performance by reordering the requests from the host adapter to minimize head switching and seeking. For example, the host adapter may ask for the following data in the following order: LBA 0, 1, 101, 102, 5, 6 (LBA = logical block address, or a byte of data). Without tagged queuing, the drive would seek to LBA 0, transfer bytes 0, then 1, then seek to 101, transfer 101 and 102, then seek back to LBA 5, transfer 5, then 6. If tagged queuing were enabled, the drive would seek to LBA 0; transfer bytes 0, then 1, 5, and 6; then seek to 101, transferring 101 and 102. At this point all the data would be transferred. Seeking on a disk drive takes a relatively long time, so reducing seeks really speeds up performance.

Taligent

A company founded jointly by Apple and IBM in 1992. Hewlett-Packard announced in 1994 that it would buy a 15% stake in Taligent. The consortium is working on an object-oriented operating system, although various independent pieces of Taligent will likely appear to be used with other operating systems (e.g., IBM's WorkplaceOS). *Pink* is an older name for Taligent, dating back to work that Apple did before the formation of Taligent.

Tank circuit

Ensures accurate signal tracking in token ring networks and prevents degradation of the signal.

TAP

Telocator Alphanumeric Protocol. A protocol for accessing pagers and transferring messages to them.

TAPI

Telephone application program interface. An *API* devised by Intel and Microsoft that allows independent developers to provide applications that integrate computing and telephony. See *CTI*.

TAT

TransAtlantic Telephone. Generic name for the various undersea cables that connect the United States and Europe. The first of the genre, TAT-1 carried 36 phone circuits between Nova Scotia and Scotland. At the time (1956) this was considered to be adequate capacity for many years. Subsequent TATs have opted for optic fiber rather than copper cables—TAT-8 was the first optic link. It was installed in 1988 and carried 8,000 circuits. The series reached TAT-13 in 1997 with a ring of cable segments connecting France, the United Kingdom, and the United States. It carries 300,000 voice circuits.

Tcl

Tool command language (pronounced "tickle"). This is designed to be a simple scripting language that can readily be extended. Tcl provides a basic set of programming capabilities (such as procedures and control structures) along with a large number of string, list, and file manipulation facilities.

TCP

Transmission control protocol. The most common transport layer protocol used on Ethernet and the Internet. It was developed by *DARPA*. TCP is built on top of the *IP* and the two are nearly always seen in combination as TCP/IP (TCP over IP). The TCP element adds reliable communication, flow control, multiplexing, and connection-oriented communication. It provides full-duplex, process-to-process connections. Defined in *RFCI* 793, TCP is connection oriented (unlike IP) and stream oriented (in contrast with UDP).

TCP/IP

Transmission control protocol/Internet protocol. The set of data communication standards adopted, initially on the Internet, for interconnection of dissimilar networks and computing systems. A de facto standard across virtually all data networks, it is commonly used over *X.25* and *Ethernet* wiring. TCP/IP operates at the third and fourth layers of the *ISO* seven-layer model (network and transport, respectively).

TDM

Time-division multiplexing. A method of combining a number of signals on a single bearer by allocating each one a predetermined time slot of the bearer. For example, a 2-Mbps bearer is split into 30 time slots, each of which is capable of supporting a 64-Kbps signal (the spare bandwidth is used for signaling and control of the 2-Mbps link). The bits from the 30 signals being carried are interleaved, transmitted, and then restored.

TDMA

Time-division multiple access. A technique for transmitting multiple data channels over a single link. The link is divided into separate time slots, with one or more allocated to each channel carried. Many of the current cellphone standards (e.g., *GSM*, *PDC*, *TACS*) are based on TDMA in preference to the promising but immature *CDMA*.

TE

Terminating equipment. The equipment (e.g., *multiplexer*) required to provide a connection point for one circuit.

TE1

Terminal Equipment 1, connecting ISDN through a 4-wire twisted pair link to NT2.

TE2

Terminal Equipment 2, connecting ISDN through a standard physical layer interface (e.g., V.24 or V.35) to a Terminal Adaptor.

Telco

Telephone company. Refers to the local, regional, or national telephone company that owns and operates lines to customer locations. See also *PTT* and *common carrier*.

Telecommunications

The general term given to the means of communicating information over a distance by electrical and electromagnetic methods. The transmission and reception of information by any kind of electromagnetic system.

Telecommuting

The practice of working at home and communicating with fellow workers over the phone, typically with a computer and modem. Telecommuting saves the employee getting to and from work and saves the employer from supplying support services such as heating and cleaning, but it can also deprive the worker of social contact and support. See also *groupware*.

Telematics

General term for computers connected over a distance. It is also the name of a well-known supplier of data communications equipment.

Telephony

Communication, often two-way, of spoken information by means of electrical signals carried by wires or radio waves. The term was used to indicate transmission of the voice as distinguished from telegraphy (done in Morse code). Telephone calls are supported by a global switching network that is well suited to the characteristics of voice (e.g., intolerance of delay, tolerance of signal corruption).

Teleworking

Using computing and communication technology to work away from an office.

Telnet

A *TCP/IP*-based application that allows connection to a remote computer. It allows a user at one site to interact with applications at other sites as if the user's terminal were local. Telnet is available as a standard package on most *PCs*.

It is defined in *RFC* 854 and has been extended with options by many other RFCs.

Terabyte

A huge amount of data, equal to 1,024 GB.

Terminal emulator

A program that allows a computer to act like a (particular brand of) terminal, such as a VT100. The computer thus appears as a terminal to the host computer and accepts the same escape sequences for functions such as cursor positioning and clearing the screen. xterm is a terminal emulator for the *X Window System*.

Terminal server

A device that connects many terminals (serial lines) to a local-area network through one network connection. A terminal server can also connect many network users to its asynchronous ports for dial-out capabilities and printer access.

Termination

A physical requirement of the SCSI bus. The first and last devices on the SCSI bus must have terminating resistors installed, and the devices in the middle of the bus must have terminating resistors removed.

Test coverage

A measure of how thoroughly a test suite exercises a program. This will typically involve collecting information about which parts of a program are actually executed when running the test suite in order to identify branches of conditional statements that have not been taken. The standard *Unix* tool for this is tcov, which annotates *C* or *FORTRAN* source with the results of a test coverage analysis.

Test data

Data used to test a program or flowchart. As well as the data, the expected results are specified.

Testing

The process of running a system with test data to check that it satisfies the specification.

Test plan

A formal set of usage scenarios that describe normal and abnormal dialogs that must be validated before new or modified software may be released.

TFTP

Trivial File Transfer Protocol. An even simpler protocol for file transfer on the Internet than FTP.

Third-generation language

A traditional high-level programming language. Eventually it was realized that assembly code was too close to the machine, so languages were designed to express the program in a form easier for a human to understand. FORTRAN, *ALGOL*, and *COBOL* are early examples of this sort of language.

Most of the modern languages (*BASIC*, *C*, C++) are third generation.

Third-party assessments

Assessments of organizations undertaken by an independent certification body or similar organization.

Throughput

A way of measuring the speed at which a system, computer, or link can accept, handle, and output information. It gives some overall efficiency, quality, and performance rating of a communications link and its associated software and protocols. Throughput is usually assessed in terms of the amount of useful information that can be carried in a certain time. Specifically:

1. The rate at which a processor can work, expressed in instructions per second or jobs per hour or some other unit of performance;
2. The amount of data that a communications channel can carry, usually in bits per second.

TickIT

A software industry quality assessment scheme initiated in the United Kingdom to provide a formal basis for capability assessment and continuous improvement.

Tie line

Two-way transmission circuits that typically connect a *PBX* in one location directly to a PBX in another. Tie lines are normally arranged for two-way calling. Calls from an extension at one location can be placed to an extension at the distant location by dialing a short access number. In most cases, this type of circuit is terminated with a four-wire analog local loop on both ends and uses multifrequency (MF) signaling. Tie lines can be used to support voice and/or data.

TIFF

Tag image file format. A method of importing graphics files into desktop publishing programs. See also *bitmap* and *pixel.*

Time-of-day routing

Routing calls based on the time the call originates (e.g., direct morning calls to East Coast operators and afternoon calls to West Coast operators).

TINA Telecommunications Information Networking Architecture. A consortium composed of leading companies in telecommunications that have come together to address the convergence of computing and telecommunications. It aims to assess and assimilate some of the emerging methods and technologies from the computing sector (e.g., client/server, object orientation) that are expected to have an impact on the telecommunications industry.

Tix A set of extensions to Tk.

Tk A graphical extension to Tcl which allows a Motif-like look and feel graphical user interface to be constructed from the Tcl scripts.

TLID Terminated Line Identity presentation and restriction. When you receive an ISDN call, your identity is sent back to the caller in the form of a TLID. Customers can request that their identities (telephone numbers) are not released to ISDN customers that are calling them.

TLV Type-length-value. A self-defining datastream.

TMN Telecommunications Management Network, a framework developed by the Network Management Forum that provides a basis for all management system functions, including those related to element, network, service, business, and resource management. Similar in scope to ONA.

TMP Theoretical midpoint. The theoretical halfway point that divides an international private line circuit into its respective U.S. and foreign halves. A U.S. records carrier is responsible for the U.S. portion of service and a foreign records carrier assumes responsibility for service to the foreign half.

TNF Third Normal Form. A technique used in data analysis and database design to reduce the complexity of the data model (and hence the amount of superfluous data that are held).

Toasternet A low-cost, low-tech, publicly accessible local community network. This is probably an extension of the term "toaster," used to mean a small, cheap, slow computer.

Token

A control information frame, possession of which grants a network device the right to transmit.

Token ring

A computer local-area network arbitration scheme in which conflicts in the transmission of messages are avoided by the granting of "tokens," which give permission to send. Token ring originated as an IBM *LAN* protocol that uses a ring-shaped network topology. Token ring has speeds of 4 and 16 Mbps. The token—a distinguishing packet transferred from machine to machine—constrains access, since only the machine that is in control of the token is able to transmit. A station keeps the token while transmitting a message, if it has a message to transmit, and then passes it on to the next station. The term is often used to refer to the IEEE 802.5 token ring standard, which is the most common type of token ring.

Tool

Any artifact that can be used to amplify the power of a human being in developing software. The term usually refers to a tangible object such as a terminal or a program (e.g., a test-case generator).

Toolkit

Collection of software components that enables certain types of applications to be built. For example, the Motif Toolkit.

TOP

Technical/office protocol. An applications layer network application and protocol stack for office automation developed by Boeing that follows the OSI model. This protocol is very similar to *MAP*, except at the lowest levels, where it uses *Ethernet* (*IEEE 802.3*) rather than token bus (*IEEE 802.4*).

Topology

A description of the shape of a network, such as star, bus, and ring. It can also be a template or pattern for the possible logical connections onto a network.

TP

Transaction processing. Controlling the rate of inquiries to a database. Specialist software—known as a TP monitor—allows a potential bottleneck to be managed.

TPC-A
A recognized benchmark for transaction processing, from the Transaction Processing Council. It is a relatively simple credit/debit benchmark. Other TPC benchmarks include TPC-B and TPC-C, the latter being a "heavier" OLTP benchmark related directly to user concerns.

TP monitor
A piece of software (often large and complex) that supports TP. Systems such as Encina and Tuxedo are widely used TP monitors.

TPON
Telephony over passive optical network. A technique for the delivery of a number of telephone circuits from a single optical fiber.

TQM
Total quality management. Continual incremental improvement in quality. A notion first put forward by W. E. Deming, widely adopted throughout Japanese industry and, somewhat later on, U.S. and European organizations.

Trace
A means of checking the logic of a program by inserting statements that cause the values of variables or other information to be printed out as the program is executed.

Trading
Matching requests for services to appropriate suppliers of those services based on some constraints. The concept of a trader, similar to that of a broker, was put forward by *ANSA*.

Traffic
A measure of the activity on a network or an individual circuit.

Traffic engineering
The process or organization responsible for monitoring historical network use statistics, anticipating growth trends, and planning, designing, and implementing network facilities.

Transaction
A single atomic unit of processing. It is usually a single small "parcel" of work that should either succeed entirely or fail entirely. Each transaction must be treated in a coherent and reliable way independent of other transactions.

Transaction processing
Abbreviated *TP*. It was originally a term that mainly applied to technology concerned with controlling

the rate of inquiries to a database. Specialist software—known as a TP monitor—allowed potential bottlenecks to be managed. It is now more widely applied to systems supporting the *ACID* properties.

Transformation
A change of one aspect or form of software into another form (e.g., the transformation of specifications for a program into the design of a program that fulfills the specifications).

Transit network
A network that passes traffic between other networks in addition to carrying traffic for its own hosts. It must have paths to at least two other networks.

Translation bridging
Bridging between networks with dissimilar MAC sublayer protocols.

Transmission
The act of transmitting a signal by electrical/electromagnetic means over a communications channel.

Transmission mode
Classification based on (1) data flow (*simplex*, *half duplex*, *full duplex*), (2) physical connection (parallel, serial), and (3) timing (*asynchronous*, *synchronous*).

Transparency
Distribution transparencies provide the ability for some of the distributed aspects of a system to be hidden from users. For example, location transparency may allow a user to access remote resources in exactly the same way as local ones.

Transparent bridging
Bridging scheme preferred by Ethernet and IEEE 802.3 networks in which bridges pass frames along one hop at a time based on tables associating end nodes with bridge ports.

Transport layer
The middle layer in the ISO seven-layer model. It determines how to use the network layer to provide a virtual error-free, point-to-point connection so that host A can send messages to host B and they will arrive uncorrupted and in the correct order. Establishes and dissolves connections between hosts. An example is *TCP*.

Transputer
A family of microprocessors developed by the Bristol, U.K.–based company, Inmos. The transputer featured interprocessor links, programmable in *Occam*.

Trap

A trap is an unsolicited (device initiated) message. The contents of the message might be simply informational, but it is mostly used to report real-time alarm information. Since a trap is usually a UDP datagram, reliance on them to inform you of network problems (i.e., passive network monitoring) is not wise.

Trapdoor function

A function that is easy to compute, but whose inverse is very difficult to compute. Such functions are good things, with important applications in cryptography, specifically in public-key cryptography.

Traveling salesman problem

A famous problem with a variety of solutions of varying complexity and efficiency. It typifies many of the problems found in both networking and computing, since it is all about finding a least-cost route to complete a known itinerary. Given a set of towns and the distances between them, determine the shortest path starting from a given town, passing through all the other towns, and returning to the first town. The simplest solution (the brute force approach) generates all possible routes and takes the shortest one. This becomes impractical as the number of towns N increases, since the number of possible routes is $(N - 1)$ factorial. A more intelligent algorithm considers the shortest path to each town that can be reached in one hop, then two hops, and so on until all towns have been visited. At each stage the algorithm maintains a frontier of reachable towns along with the shortest route to each. It then expands this frontier by one hop each time.

Trigger

In general, the activation of an event-driven process. When applied to a database, it is an application-specific process invoked by a database management system as a result of a request to add, change, delete, or retrieve a data element.

Trojan horse

A malicious security-breaking program that is disguised as something benign, such as a directory listing, archiver, game, or (in one notorious 1990 case) a program to find and destroy viruses! A Trojan horse is similar to a backdoor.

Trouble ticket

Some record (originally a piece of paper) used to report and track the resolution of a network or circuit problem. Handling trouble tickets is part of the service management process, and facilities for dealing with them are built into modern network and service management tools.

TrueType

An outline font standard first developed by Apple Computer, and later embraced by Microsoft, as a competitor to Adobe's PostScript, which is still more popular.

True Voice

An AT&T offering, intended to provide faithful voice reproduction over standard telephone lines.

Trumpet

A news reader for Microsoft *Windows* using the WinSock library. There is also an MS-DOS version. Trumpet is shareware from Australia.

Trunk

A connection between network switches, usually a high-capacity one. It is typically at least a *T1* (in the United States) or an *E1* (in Europe).

Trunk group

A group of circuits of a common type that originate from the same location.

TSAPI

Telephone services application program interface. An API devised by AT&T and Novell that allows independent developers to provide applications that integrate computing and telephony. See *CTI.* It is similar to *TAPI* and is probably the more popular of the two.

TSR

Terminate and stay resident. A type of *DOS* utility that, once loaded, stays in memory and can be reactivated by pressing a certain combination of keys.

TST

Time Space Time, a method of switching allowing a number of channels to share the same physical switch by allocation to each over a short period of time.

Tunneling

Usually refers to a situation where a public network is used to connect two or more private domains so that the privacy of the overall link is maintained. Tunneling is frequently used in the construction of corporate intranets, with the public telephone network providing the connectivity between sites.

Two phase commit

Protocol for transactional coordination of updates in two or more databases.

ABCDEFGHIJKLMNOPQRSTUVWXYZABCDEFGHIJKLMNOPQRSTUVWXYZABCDEFGHIJKLMNO

U
ISDN reference point between NT1 devices and line termination.

UA
User agent. An *X.400* component of an *MHS* that assists users in the preparation, storage, and display of messages.

UART
Universal Asynchronous Receiver/Transmit chip, used to drive the communications (COM) port in personal computers.

UBR
Unspecified Bit Rate, a class of service for delay tolerant traffic such as file transfers.

UDP
User datagram protocol. An Internet standard, UDP is a connectionless protocol that, like *TCP*, is layered on top of *IP*. It provides simple and efficient, if unreliable, datagram services. It is defined in *RFC* 768. UDP is required to carry many of the more basic Internet services.

UltraSCSI
A method that enables very fast data transfer rate on the SCSI bus. The maximum UltraSCSI data transfer rates are 20 MBps on 8-bit (40 MBps for Wide SCSI host adapters).

Ultrix
A version of the *Unix* operating system (based on the *Berkeley* version) that was designed and imple-

mented by *DEC* to run on their VAX and DEC-station processors.

UML

Unified Modeling Language. The object-oriented notation adopted by the OMG and devised by Booch, Rumbaugh, and Jacobson. UML unifies the formerly disparate flavors of object-oriented notation.

UMTS

Universal Mobile Telephony Service. This is the concept that users will be able to access all of their network services irrespective of location. The goal of UMTS is to provide a network based around the individual—the individual registers his or her access device with the network rather than having to find a fixed access device.

UNI

User network interface. An *ATM* Forum specification of the interface between the ATM network and a connected ATM terminal device.

Unified notation

The planned consolidation of the object-oriented approaches of Booch and Rumbaugh in respect of presentation standards.

Uniprocessor

A computer having only a single main processor.

Unix

One of the most important of modern operating systems. It was developed by a university rather than by a commercial vendor and gained popularity with practitioners because of its efficiency. Unix is now owned by *X/Open* and is available from many vendors. It is the basis of open systems—the unbundling of computer software that allows interchangeability between various commercial products.

Unix International

A consortium including Sun, AT&T, and others formed to promote an open environment based on Unix System V, including the Open Look windowing system.

UNMA

Unified Network Management Architecture, a proprietary framework for managing networks, devised some years ago in AT&T.

U-plane

User plane. Part of the ISDN parlance that refers to the information switching part of a network (i.e., the actual data carried from one end to the other).

Upload

To transfer a file from your computer to another computer, using your terminal program (e.g., Qmodem) and a transfer protocol (e.g., Zmodem).

UPT

Universal personal telephony. The idea that users can get their telephony services wherever they are by identifying themselves to the network through some form of user ID. The concept has been largely overtaken by the rapid rise of mobile communications, which provide a more convenient means of getting ubiquitous service.

URL

Uniform resource locator. A standard for locating an object on the Internet and most widely known as the form of address for pages on the *World Wide Web*. Typical URLs take the form http://www.myname.com/, ftp://archive.ic.ac/ john, or telnet://jungle.com. The part before the first colon specifies the access scheme or protocol. The part after the colon is interpreted according to the access scheme. In general, two slashes after the colon indicate a host name.

USB

Universal Serial Bus, an I/O bus standard jointly developed by Intel, Compaq, DEC, IBM, Microsoft, NEC, and NT. Aims to eliminate separate connectors for peripherals (e.g., printers, modems, joysticks, scanners, muse devices, keyboards).

Use case

A method for capturing the functional requirements of computer systems. In essence, it works by working with a customer to decide what business functions a system must support and then documenting a narrative scenario (the use case) for each of those functions.

Usenet

A distributed bulletin board system supported mainly by *Unix* machines. It is probably the largest decentralized information utility in the world, encompassing government agencies, universities, schools, businesses, and hobbyists. Usenet has well over 1,200 host machines, and incorporates the equivalent of several thousand paper pages of new technical articles, news, discussion, and opinion every day. To join in, users need a news reader such as xrn or one of the newer browsers (e.g., Netscape 2) that incorporate mail and news facilities.

Utility program

A systems program designed to perform a common task such as transferring data from one storage device to another or editing text.

UTP

Unshielded twisted pair. Normal telephone wire used for computer-to-computer communications, usually as Ethernet or localtalk cabling. It is much cheaper than standard "full-spec" *Ethernet* cable.

UUCP

Unix-to-Unix Copy or Unix–Unix Communication Protocol. A *Unix* utility program and protocol that allows one Unix system to send files to another via a serial line. It is a fairly basic facility and has generally been superseded by *FTP* or rcp for file transfer, and *SMTP* for electronic mail.

UUdecode

A Unix program to convert the ASCII output of UUencode back to the original binary form.

UUencode

Unix program for encoding binary data as ASCII. UUencode was originally used with UUCP to transfer binary files over serial transmission links. It is still widely used for sending binary files by e-mail and posting to Usenet newsgroups.

UUID

Universally unique identifier. A naming scheme that can be used to ensure that objects in distributed systems are not confused with each other.

V.21

An *ITU-T* modem protocol for 300-bps two-wire, full-duplex communications using frequency-shift keying modulation.

V.22

An *ITU-T* modem protocol that allowed data rates of 1,200 bits per second. The extension of V.22 (V.22bis) doubles the rate to 2.4 Kbps.

V.24

An *ITU-T* standard for the interface between a terminal and a piece of communication equipment (e.g., a PC to modem connection). Usually implemented as a 25-pin connector, similar to *RS232.*

V.3

The ITU-T standard that defines the International Alphabet No. 5.

V.34

An *ITU-T* standard for the transmission of data over the public network. It is intended for data transmission up to 28.8 Kbps (a rate used by the type of high-speed modems used by many home users for *Internet* access).

V.35

An *ITU-T* standard for the transmission of data over the public network. Intended for data transmission at 48 Kbps. It uses the 34-pin V.34 connector, which is similar to RS-422/RS-449.

VADSL

Very high-rate ADSL (also known as VDSL and VHDSL) that uses the same transmission techniques as ADSL but at higher speed (and with shorter line lengths). It offers 10–25 Mbps over a copper drop of a few hundreds of meters.

Validation

Making sure that the right thing is being done. The formal process of reviewing or examining something (e.g., a requirements specification, user interface, plan, budget) and confirming that is in line with expectations. It is the complementary process to verification.

Value-added reseller

A company that sells something made by another company with the addition of extra components (e.g., specialist software on top of a proprietary computer).

VAN

Value-added network. A communication network that provides features other than transmission of information. The value added is usually provided by the translation of one type of computer signal to another type of computer signal. The term sometimes refers to packet-switched networks with protocol conversion (that allow dissimilar systems to work together).

Vaporware

A disparaging term used when a proposed or advertised product (usually software-based) is not backed up with a working product.

Variant

A standard product with some level of customization. It is usually a software delivery that has site- or user-specific features. Effective configuration management requires control over variants as well as versions. The former tend to be customer-specific and the latter are usually determined by the plans of the developers.

Vaxen

The attractive and whimsical, if incorrect, plural of VAX—the machines from Digital Equipment Corporation that were once the mainstay of many academic and commercial computing environments.

VBR

Variable Bit Rate. Usually refers to real-time services such as voice or video where the packets that carry the information are only generated when necessary—none during silence, or still scenes in video.

VCC

Virtual Channel Connection. Defined by a series of VCs logically assigned to make an end-to-end link.

VCI

Virtual Connection Identifier. A 16-bit identifier having only local significance on the link between ATM nodes.

VCI

Virtual channel identifier. Identifies, in an *ATM* network, the individual channels carried across a given path. Each of the channels may carry a different service (e.g., a voice link, a *frame relay* connection) and have its own traffic contract.

VC-n

Virtual Container, the SDH container for a PDH bit rate (including bit-stuffing) plus the path overhead (POH).

VDM

Vienna Development Method. One of the first methods of software development to be based on mathematics. VDM provides a route to proving software designs to be correct and is hence used for safety- and security-critical applications. It requires considerable expertise to apply correctly.

VDSL

Very high-rate digital subscriber line. A faster variant of *ADSL*, with more limited reach. VDSL offers speeds in the 25- to 51-Mbps range (*ATM* transport rates) with a 500-Kbps to 2-Mbps return path. The range is limited to a few hundred meters of copper wire, but the technology does enable services such as high-definition television to be provided without fibers having to be laid.

Vendor-independent

Hardware or software that will work with hardware and software manufactured by different vendors—the opposite of proprietary.

Verification

Making sure that something is being done right. Verification is the process of searching for and eliminating errors and is the complement of *validation*.

Veronica

Very easy rodent-oriented netwide index to computerized archives. An *Internet* facility that is accessed through Gopher. It allows a user to carry out a keyword search on Gopher titles.

Version Generally denotes a particular collection of items that make up an identifiable product or baseline. The term is commonly used in software development to define a compatible set of components. The control over versions is part of configuration management.

VESA Video Electronics Standards Association. The VESA bus is a PC bus, that will work on any existing bus—ISA, EISA, microchannel.

VFS Virtual File Storage. An intermediate format used for data in transit from one system to another system. It is used as the transit format for File Transfer, Access, and Management (FTAM) and provides a set of common file operations that all FTAM systems understand (such as copy or delete).

VGA Video graphics adapter. A display standard for IBM PCs that has 640 × 480 pixels in 16 colors and a 4:3 aspect ratio. It is usually superseded by higher resolution standards (such as super VGA, or SVGA).

Video compression A technique that compresses sequences of images, relying on the fact that there are usually only small changes from one "frame" to the next. Encoding is only required for the starting frame and for a sequence of differences between frames. *MPEG* and H.261 are examples of standards in this area.

VINES Virtual Networking System. A *LAN* product from Banyan, which is similar in function to Novell's *NetWare*.

Virtual circuit A transmission path through an *X.25* public switched data network (*PSDN*) established by exchange of setup messages between source and destination terminals. There is no physical end-to-end path reserved for a call; information gets from source to destination over whatever resources are available.

Virtual device A module allocated to an application by an operating system or network operating system instead of a real or physical one. The user can then use a computer facility (keyboard, memory, disk, or port) as though it were really present. In fact, only the operating system has access to the real device.

Virtual machine

An assumed resource that exists as a definition rather than as a real machine. The concept of a virtual machine is useful in that it gives developers a target for their applications. They can write for the virtual machine and reasonably expect the translations and additions required by a range of real machines to be readily available.

Virtual path

The location of a file or directory on a particular server, as seen by a remote client accessing it via the *World Wide Web* (or any similar distributed document service). A virtual path provides access to file soutside the default directory. It appears in the form ".../somename/...," where "somename" is replaced with an actual path configured by the administrator.

Virtual PoP

Virtual point of presence. An access point for users to connect to an Internet access provider that is not operated by the provider. Users are charged by the telephone company for the call to the virtual PoP, which relays their call via some third-party circuit to the Internet provider's central location. This contrasts with a physical PoP, which is operated by the Internet provider itself. The advantage of a virtual PoP is that the provider can keep all its modems in one location, thus improving availability and maintenance, but users do not have to pay long-distance call charges to that point.

Virtual route

SNA terminology for virtual circuit. A logical connection between subarea nodes that is physically realized as a particular explicit route.

Virtual team

A group of people working together on the same project who are physically separate, their only link being via a network and computer screens.

Virus

A program passed over networks that has potentially destructive effects once it arrives. Packages such as Virus Guard are in common use to prevent infection from hostile visitors. Unlike a *worm*, a virus cannot infect other computers without assistance. It is usually propagated by people who share programs with their friends and colleagues.

Visual Basic

One of a number of programming languages that presents the user with a range of on-screen graphical objects. Widely used for producing attractive prototypes very quickly. Visual *C* and Visual C++ are similar to look at, the difference being the underlying code used in the application that is produced.

VME

Visa-Mastercard-Europay. A proposal for an international standard for smart cards with stored value functions. Also, an operating system from ICL that was popular in the 1980s.

VMS

Virtual machine system. An operating system from Digital that was popularized in the VAX range of computers.

VoD

Video on Demand. A service that allows access to videos over the telephone network. Access is over standard links, using high speed transmission to cope with the volume of data being sent. The videos are held on network connected databases in digital form.

Voice mail

An automatic answering service with the ability to record a message. Unlike simple answering machines, voice mail uses a programmable computer system to provide a variety of user configurable options (e.g., divert).

VOIP

Voice Over IP. The configuration of an IP network, designed to carry data traffic on a best-effort basis, to carry voice calls. This entails the deployment of mechanisms such as Diffserv that ensure the timely delivery of packets containing voice samples.

VP

Virtual Path. In ATM, a link that contains several virtual connections (VCs). A VP is always routed as a whole by ATM cross-connects, so VCs within it can be allocated to individual connections on that path. A feature of VPs is that that they are elastic-sided, with their bandwidth varying as the VC bandwidths vary.

VPI

Virtual path identifier. Identity of a connection that is set up in an *ATM* network on a subscription basis by the network provider. It provides a route between two sites and enables bulk cell switching between those sites.

A number of channels can be supported within one virtual path. Each of these channels is tagged with a *VCI*.

VPN

Virtual private network. The provision of facilities that are normally associated with a private network via a public network. In effect, some portion of the public network is configured to serve a closed group of users.

VRML

Virtual Reality Markup Language. An extension of the *HTML* concept into virtual reality. VRML provides a language for coding virtual reality images that can be accessed over a network by anyone with a compatible browser.

VSAM

Virtual sequential access method. A disk file storage scheme used in IBM's OS/360 operating system. VSAM improved access time and reduced the need for reorganization of certain types of randomly accessed disk files.

VSAT

Very-small-aperture terminals. Small satellite dishes used to communicate via satellites. They provide an easy-to-install digital connection and can interwork with all of the common networking protocols. They provide bandwidth up to 128K and are well suited to applications that require the same information to be broadcast to many remote sites. Despite being expensive compared to terrestrial options, they are also a viable alternative to a router for hostile or awkward environments.

V series

The set of *ITU-T* recommendations covering data transmission over telephone circuits.

VTAM

Virtual telecommunications access method. A data communications access method that is compatible with IBM *SNA*.

ABCDEFGHIJKLMNOPQRSTUVWXYZABCDEFGHIJKLMNOPQRSTUVWXYZABCDEFGHIJKLMNO

WABI

Windows application binary interface. A software package that emulates the popular Microsoft Windows system, but which runs under the X Window System (usually on a Unix workstation rather than a PC). It is a specific example of an *ABI*.

WAIS

Wide-area information server. One of a number of *Internet* utilities, used for public database text searching. The search returns a list of documents, ranked according to the frequency of occurrence of the keyword(s) used in the search.

The client can retrieve text or multimedia documents stored on the server. Other information retrieval systems include *Archie*, *gopher*, and the *World Wide Web*.

Walkthrough

A peer review of a system design, code, or detailed technical plan that has the aim of identifying errors as early as possible and learning from other people's experience. It is widely used as a means of assuring quality during software development.

WAN

Wide-area network. A network (usually one provided by a national public operator) that spans a long distance (e.g., several miles, beyond the reach of an *Ethernet*) and serves a large number of people (often the general public).

WARC World Authority on Radio Communications. The body responsible for the allocation of the radio spectrum. Rulings from WARC are of vital interest to the wide range of communications, network, and service providers intent on delivering their wares via radio.

Waterfall The "classical" software life cycle, so named because the chart used to portray the phases of development suggests a waterfall.

WATTC World Administrative Telegraph and Telephone Conference. An *ITU*-organized meeting set up with the aim of agreeing how international telecommunications should be handled.

WDM Wavelength-division multiplexing. A sophisticated technique for the transmission of very high data rates over optical fibers. As the name indicates, the idea is to have two or more light sources, each transmitting at a different wavelength.

Webmaster A nominated keeper of a set of *World Wide Web* pages. The webmaster is often the system administrator for the server providing the pages.

WFQ Weighted Fair Queuing. In an IP network, this provides the capability to expedite the handling of high priority, low delay traffic, while equitably sharing the remaining bandwidth between lower priority traffic.

White line skipping A compression technique used with facsimile systems whereby blank lines are not coded.

White pages A computer network directory service, rather like the telephone directory, for locating individuals on the Internet by name. The *Internet* supports several databases that contain basic information about users, such as electronic mail addresses, telephone numbers, and postal addresses, and these databases can be searched to get information about particular individuals. Utilities such as *whois* and *finger* allow a user to carry out a search.

Whois

An *Internet* program that allows users to query the identity of someone connected to their system. Similar in function to *finger*.

Wideband

A communications circuit that is wider than narrowband but not as wide as *broadband*. It is usually taken to lie in a range between 64 or 144 Kbps and 1.544 or 2 Mbps.

Wideband CDMA

Mainstream air interface solution for third generation wireless networks.

Wide SCSI

Provides for performance and compatibility enhancements to SCSI-1 by adding a 16- or 32-bit data path. Combined with Fast SCSI, this can result in SCSI bus data transfer rates of 20 MBps (with a 16-bit bus) or 40 MBps (with a 32-bit bus).

WIMP

Windows, icons, mouse (or menu), pull-down menus. The style of graphical user interface popularized by the Apple Macintosh and now commonplace in, for example, the personal computer and the *X Window System*.

Win32 API

A 32-bit application programming interface for both Windows for MS-DOS and Windows NT. It updates earlier versions of the Windows API with sophisticated operating system capabilities, security, and API routines for displaying text-based applications in a window.

Window

A flow control mechanism whose size determines the number of data units that can be sent before an acknowledgment of receipt is needed and before more can be transmitted.

Windows

A way of displaying information on a screen so that users can do the equivalent of looking at several pieces of paper at once. Each window can be manipulated for closer examination or amendment.

This technique allows the user to look at two files at once or even to run more than one program simultaneously. Examples of windows systems are the X Window System and proprietary systems such as Windows 3.1 on the PC, NeWS on Sun workstations, and the Macintosh.

Windows 95

Microsoft's successor to its Windows 3.1. It was introduced in 1995.

Windows NT

The NT stands for new technology. This is Microsoft's 32-bit operating system targeted at the workstation and corporate network market. Unlike Windows 3.1, which was a graphical environment that ran on top of DOS, Windows NT is a complete operating system (with features like true multithreading, built-in networking, and security). From the user's point of view, however, it looks much the same as Windows 3.1.

Winsock driver

A widely used package that mediates between Windows and the TCP/IP protocols.

WLL

Wireless local loop. An alternative to using copper to provide access to a telephone network. WLL tends to be deployed in difficult or hostile environments and allows rapid provision of service.

Word

Inside a computer, the sequence of bits or characters that is treated as a single unit. The length of a word is usually determined by the internal design of the computer.

Common word sizes are 32 bits (for powerful mainframes such as the IBM System/370), 16 bits (used in many *PCs*), and 8 bits (used when simplicity and speed are paramount, as in *RISC*).

Workstation

A networked personal computing device with more power than a typical PC. It is usually a *Unix* machine capable of running several tasks at the same time. The term is often used to refer to a computing system that is more powerful than a simple personal computer.

World numbering plan

An *ITU* plan that divides the world into nine zones, with each one having a different digit allocated to it. This digit forms the first number of the country code for each country in that zone.

World Wide Web

Also known as WWW and W3, the Internet-based distributed information retrieval system that uses hypertext to link multimedia documents. This makes the relationship of information that is common between documents easily accessible and completely independent of physical location.

WWW is a client/server system. The client software takes the form of a *browser* that allows the user to easily navigate the information online. Well-known browsers are Netscape and Mosaic. A huge amount of information can be found on WWW servers.

On the WWW, everything (documents, menus, indexes) is represented to the user as a hypertext object in *HTML* format. Hypertext links refer to other documents with their *URLs*, which can refer to local or remote resources accessible via *FTP*, *Gopher*, *Telnet*, or news, as well as those available via the HTTP protocol used to transfer hypertext documents.

Worm

A computer program that replicates itself—a form of virus. The *Internet* worm was probably the most famous; it successfully, and accidentally, duplicated itself across the entire system.

WRED

Weighted Random Early Detect. This is a mechanism that combines IP precedence and Random Early Detect (RED) capabilities to provide differentiated performance characteristics for different classes of service.

X.1

The ITU-T standard that describes international user classes of service in public data networks.

X.3

The *ITU-T* standard that specifies the basic functions and capabilities of a *PAD*, a device that allows non-*X.25* terminals to connect over an X.25 network.

X.11

An ITU-T standard that is a common abbreviation for X-Windows, referring to version 11.

X.21

An *ITU-T* standard interface for connection of a data terminal to a network. X.21 is similar to the familiar *RS-232* interface, except that it uses a smaller connector and has two wires for call control.

X.25

A widely used *ITU-T* standard protocol suite for packet switching, which is also approved by *ISO*. X.25 defines standard physical layer, link layer, and network layers (i.e., ISO layers 1 to 3). It was originally developed to describe how data should pass into and out of a public data communications networks. X.25 is the dominant packet data standard. Many X.25 products are available on the market, and networks have been installed all over the world. There are a number of other ITU-T recommendations that relate to packet switching: *X.3*, *X.28*,

X.29, and *X.75*. These are often used in conjunction with X.25.

X.28

The *ITU-T* standard that specifies how you control a *PAD* from a terminal. With X.29, it allows character mode terminals to connect to an X.25 network.

X.29

The *ITU-T* standard that specifies procedures for the exchange of control information and user data between a *PAD* and a remote packet-mode terminal.

X.75

The *ITU-T* standard that specifies the protocols to be used for communication between two public switched data networks. It can be thought of as the internal network transport for *X.25* packets.

X.110

The ITU-T standard that covers the routing principles for international public data services through switched public data networks of similar type.

X.121

The ITU-T standard that contains the numbering plan for public data networks.

X.400

A set of *ITU-T* communication standards covering electronic mail services provided by data networks. It is widely used in Europe and Canada as the basis for e-mail services.

X.400 consists of a store-and-forward *MHS* specification that allows for the electronic exchange of text as well as other electronic data such as graphics and fax. It provides a means for suppliers to interwork between different electronic mail systems.

X.400 has several protocols, which are defined to allow the reliable transfer of information between two main X.400 system elements—*UAs* and *MTAs*. X.400 addresses tend to be very long compared to those used across the *Internet* (as defined in *RFC 822)*.

X.435

The ITU-T standard that covers the transmission of *EDI* messages over X.400 mail systems.

X.500

A set of ITU-T standards covering electronic directory services such as white pages and *whois*. X.500 is designed to permit applications such as electronic mail to access information that can be either central or distributed.

X.711 ISO IEC 9596-1 standard on Information Technology—Open Systems Interconnection—Common Management Information Protocol Specification, CMIP.

X.720 ISO IEC 10165-1 standard on Information Technology—Open Systems Interconnection—structure of management information—management information model.

X.800 ITU-T standard for security architectures.

XA An extension of IBM's *MVS* operating system, intended for very large mainframe computers. Usually associated with high-volume transaction processing applications.

XDR External data representation. The "on-the-wire" data format used by Sun *RPC.*

XFN X/Open federated naming standard. A naming standard used in distributed computing systems that follows on from *NIS+.*

XML Extended Markup Language. A more powerful, and general, successor to HTML.

XNS Xerox network services. A layered set of data communication protocols from the Xerox Corporation. XNS was an early solution in distributed systems that set the scene for subsequent products.

X-off The character (control-U) that is used to restart the flow of characters.

X-on A character (control-S) that is used for flow control. It is used to stop the flow of characters during a data transmission.

X/Open An international industry standards consortium that develops detailed system specifications by drawing on available standards. X/Open owns the *Unix* trademark and thereby provides a common reference for the industry variants (e.g., HP-UX, *AIX* from IBM, Solaris from Sun). X/Open also defines the X/Open Common Applications Environment, which aims to provide applications portability.

XPG3

Version 3 of the *X/Open* Portability Guide. A comprehensive guide to producing portable applications. It contains preferred system calls—those most likely to be consistently implemented on the widest range of vendor equipment.

X protocol

A standard protocol used by both clients and servers in the X Window System. It allows the exchange of requests for window operations.

X series

The series of *ITU-T* recommendations that cover data transmission over digital circuits.

X terminal

An intelligent terminal that operates as an X server directly connected to Ethernet.

X Window System

A specification for device-independent windowing operations on bitmap display devices. It is a de facto standard supported by the X Consortium, although precise implementation details (e.g., location of scroll bar) do tend to vary according to supplier. X originated with and is used on many Unix systems.

ABCDEFGHIJKLMNOPQRSTUVWXYZABCDEFGHIJKLMNOPQRSTUVWXYZABCDEFGHIJKLMNO

YACC
Yet another compiler compiler. A standard facility on the *Unix* operating system used (with an associated utility called "lex") for generating the language parsers required to interpret textual input.

Yahoo
Yet Another Hierarchically Organized Oracle. One of the many search utilities that can be used to find specified items within the information held on the *World Wide Web*. Others include *Lycos*, Alta Vista, and Excite. Yahoo is rumored to stand for any of "yet another hierarchically officious, obstreperous, or organized oracle," or possibly it is just named after a race of brutes in Swift's *Gulliver's Travels*. Yahoo is probably the biggest hierarchical index on the *World Wide Web*.

Yellow Pages
Used to be the name for Sun Microsystems' *NIS* and is still widely used.

Y modem
A file transmission protocol that sends known file sizes in *batch*. It is a successor to X modem.

Yourdon
Name of a well-known innovator in software engineering. Yourdon is the generic name given to the systematic design of software systems using data flow as the primary means of analysis.

ABCDEFGHIJKLMNOPQRSTUVWXYZABCDEFGHIJKLMNOPQRSTUVWXYZABCDEFGHIJKLMNO

Z — A mathematically formal specification language developed by the Programming Research Group at Oxford University. It is used for describing and modeling computing systems and has proved effective in uncovering subtle design oversights.

Z.100 — ITU-T standard—Specification and Description Language (SDL), widely used to describe the design and operation of telecommunications systems.

ZDL — Zero Delay Lockout. Technology designed to prevent beaconing stations on a token ring from inserting into the ring and causing faults.

Zip — A compression program from PKWare to reduce files that are to be sent over a network to a more reasonable size. It was originally popularized on MS-*DOS*, but has now spread to other operating systems.

Z modem — A file transfer protocol with error checking and crash recovery. It is a successor to *Y modem.*

Zone — A logical group of network devices on an AppleTalk system.

Zoo

Another compression program, originally developed for the Macintosh.

Zulu time

Greenwich mean time. A reference point for system time.

ABCDEFGHIJKLMNOPQRSTUVWXYZABCDEFGHIJKLMNOPQRSTUVWXYZABCDEFGHIJKLMNO

1base-5 1-Mbps transmission over twisted-pair configuration.

2-tier Two-tier *client/server* applications consist of a single client and server pairing. Often the server is a central database with most of the application code placed in the client.

3-tier Three-tier *client/server* systems partition the application code according to the type of processing it performs. This may, but does not have to be, distributed across three different physical systems.

4GL Fourth-generation language. A term usually applied to languages such as *SQL* that are designed to allow databases to be interrogated.

10base-2 A form of thin coaxial network cable, often referred to as *cheapernet*, that is used with *Ethernet* installations. The term is built up from the supported network speed (10 Mbps), the type of transmission (baseband), and the maximum transmission reach (200m).

10base-5 A thick coaxial network cable used with *Ethernet* installations. Can support segments of up to 500m at 10 Mbps.

10base-T
A twisted-pair network cable for *Ethernet* installations. At 10 Mbps, the recommended limit on segment length is 100m.

100base-FX
IEEE 802.3 physical layer specification for 100-Mbps Ethernet over two strands of fiber.

100base-T
A group of IEEE 802.3 physical layer specifications for 100-Mbps using baseband over various wiring specifications.

100base-T4
IEEE 802.3 physical layer specification for 100-Mbps Ethernet over four pairs of category 3, 4, or 5 UTP wire.

100base-TX
IEEE physical layer specification for 100-Mbps CSMA/CD over two pairs of category 5 UTP or STP wire.

100base-VG
Proposed standard now under consideration by a newly created working party of the IEEE 802 committee which is considering access methods to 100-Mbps transmission over *UTP*. Proposers include Hewlett-Packard and AT&T.

100VG-AnyLAN
Method of *LAN* transmission based on a new 100base-VG standard, now awaiting *IEEE* approval, that builds on aspects of both token ring and Ethernet to provide a new way of transmitting LAN traffic at 100 Mbps over existing *UTP* wiring.

802.3
Broadband bus networking system. It uses the *CSMA/CD* protocol. *Ethernet* has become the generic name, despite its being one trademarked version of 802.3.

802.4
Standard that governs broadband bus and broadband token bus. It is usually used in industrial applications.

802.5
Standard that governs token-ring networking systems.

802.6
Standard that governs *MANs*.

2780
A batch standard used to communicate with IBM mainframes or compatible systems.

3270

IBM's interactive communications terminal standard used to communicate with an IBM mainframe or compatible systems.

3780

A batch protocol used to communicate with an IBM mainframe or compatible systems.

Trends in Information/Technology

ABCDEFGHIJKLMNOPQRSTUVWXYZABCDEFGHIJKLMNOPQRSTUVWXYZABCDEFGHIJKLMNO

History is philosophy drawn from examples.
—Dionysius of Halicarnassus

It is much easier to assimilate information when you know the context. In spite of the rapid advances in technology over the last 30 years, there are some general trends that have emerged. Even if they do not directly help in predicting the future, they help make sense of where you are and where you are headed.

The following illustrations are grouped into a number of key areas, each with its own story to tell. Each one has an associated narrative that explains how we got here and what the key issues are. The information presented here is drawn from a very wide variety of sources. Only a very little was already neatly packaged to illustrate a specific point—most of the following has been compiled partly from published reports, partly from intelligence gathered from the workplace, and partly from observation of the industry. As a safeguard, I have discussed these trends with a number of trusted colleagues. No one took issue with the picture that emerges. The detail may be subject to interpretation or debate, but the directions are clear.

The User and the Marketplace

One change always leaves the way prepared for the introduction of another.
—Niccolo Machiavelli

We start by looking at expectations—what the users see of technology, what they expect of it, and how they would like it to evolve. The issues dealt with here are:

- User involvement with technology;
- Speed to market;
- User interaction;
- The effect of the personal computer;
- Speech recognition.

User Involvement With Technology

There is little doubt that technology has consistently moved more and more into the mainstream fabric of society. Figure 1 illustrates the increasing acceptance of personal information devices by the general public. The obvious message is that people now accept technology. They are comfortable with it. They are likely to want more, better, faster, and will expect it to work as a normal part of their environment. Some specific trends in this are that:

- The sheer volume of information carried around by each person will have to increase to satisfy expectation.
- Higher degrees of networking will be expected. People will want access to more library information, databases, and other resources.

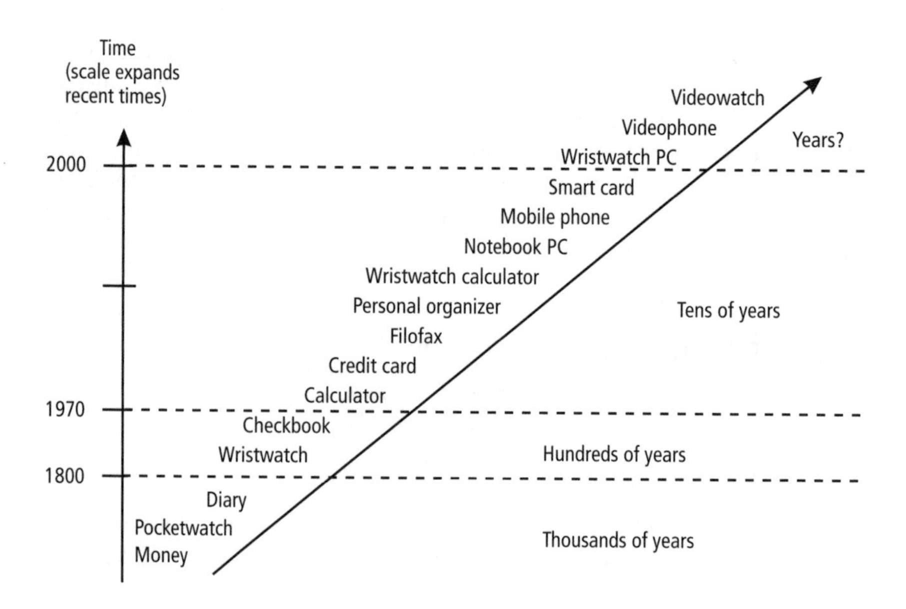

Figure 1 User involvement with technology.

- Increased portability is a consistent theme that seems set to continue. The mobile phone has blazed the trail. Other devices will have to be usable wherever the user wants to use them.
- More sophisticated interfaces will have to be deployed in order to capitalize on the above. Speech, for instance, is likely to become increasingly favored over the keyboard for driving a computer.

Speed to Market

Technology drives change, and change drives technology. In general, the pace of life has accelerated through the twentieth century. The simple message being illustrated in Figure 2 is that this is pressing technical innovation very hard.

Not only has technology change been speeding up, so has market entry and diffusion into society. Vendor strategies have changed as a result. An important aspect of this is the way that distribution is organized, with more third-party involvement by organizations that are adept at moving products quickly. In particular, we are increasingly seeing:

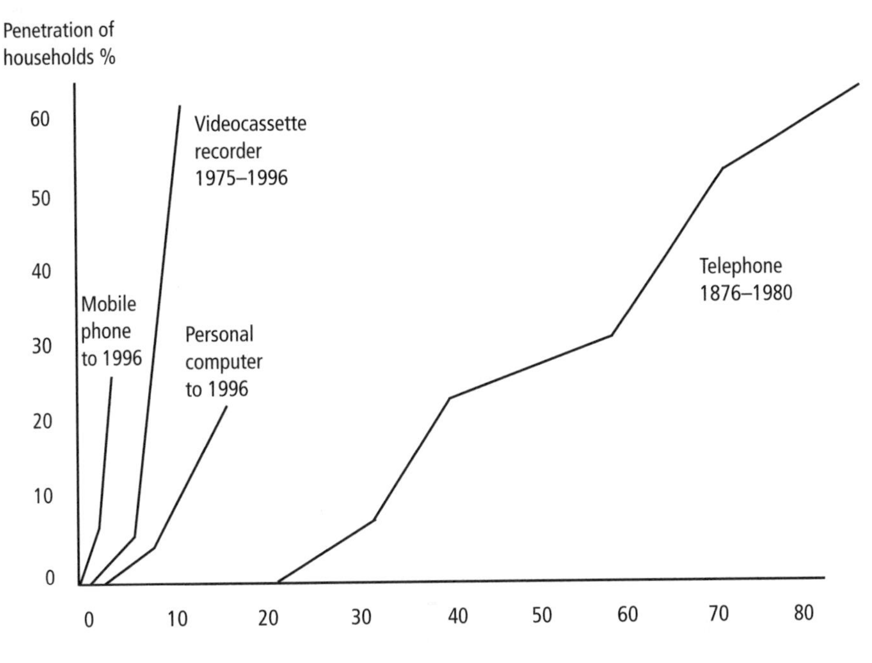

Figure 2 Speed to market.

- Direct sales taking precedence over retail operations. A trend that has been driven, to a large extent, by the Internet.
- More "bundling" of products. The convergence of telecommunications and computing is increasingly allowing products to be assembled from a set of components, leaving the seller to differentiate on service.

User Interaction

The interface between users and the technology they work with has been getting more sophisticated (see Figure 3). Once it was primitive, oriented more toward the needs of the machine than those of the human. The converse is now true, with the usability of high-tech products being a major consideration. Nowhere is this more clearly seen than in the way in which we access computer systems, or rather the services they provide. The usability of browsers such as Netscape has made the information on the Internet accessible to a huge number of people.

One of the underlying reasons for wanting straightforward access is to give more scope for controlling increasingly complex systems, but there are others. For instance, a wider population of "ordinary users" now use the computer, and they need an interface that is both intuitive and reliable.

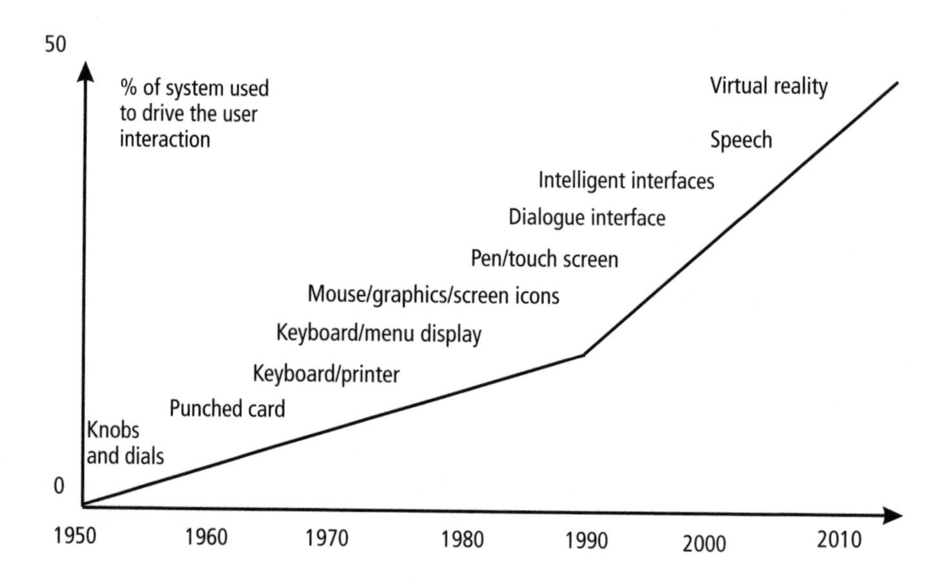

Figure 3 User interaction.

Much of the dramatic increase in computing power has been absorbed by these interfaces that have become more graphical and, with virtual reality, more apparently lifelike, allowing users to feel as if they are an integral part of the system. Inevitably, interfaces will provide yet more guidance (e.g., through the use of intelligent agents), allowing the user to focus on the principal tasks.

The Effect of the Personal Computer

The price of computing power has steadily decreased over the years. This averages out to around a factor of 10 every 10 years.

In recent years, the effect of the personal computer has promoted an even steeper decline. Looking forward from the beginning of the twenty-first century, the conventional mainframe computer will not only decrease in price per unit of power, but will also have to approach the technologies of the "alternative mainframes." That is, they will use cheaper technology and adopt a more parallel architecture.

The general point being illustrated in Figure 4 is the drift from centralized computing (as was prevalent when computer bureaus flourished as a means of sharing an expensive resource) to a more distributed arrangement

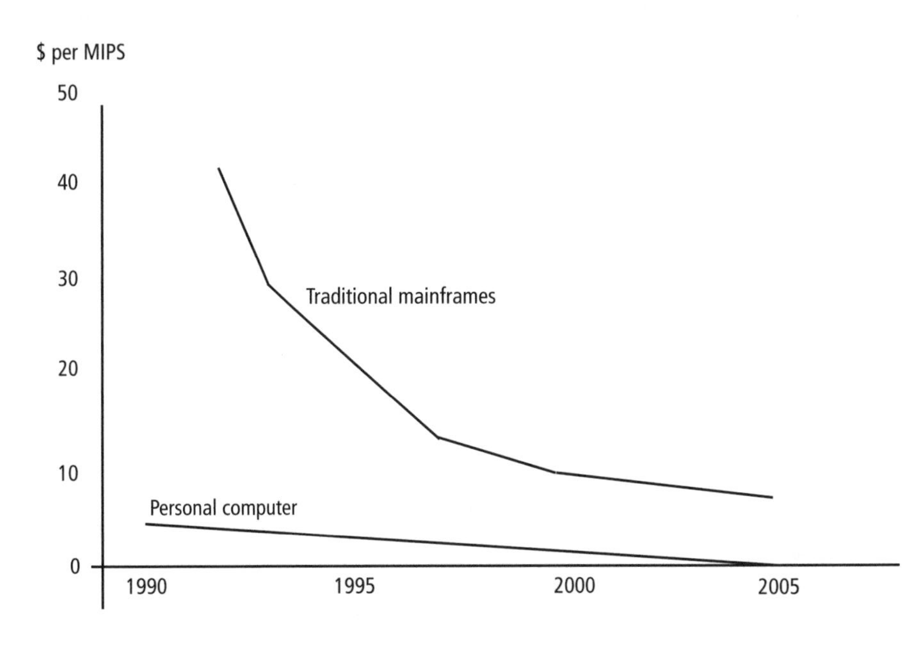

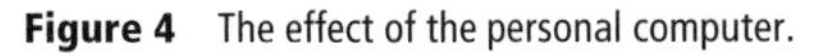
Figure 4 The effect of the personal computer.

(where networks such as the Internet and architectures such as client/server reduce entry costs to a minimum). Not only has the personal computer popularized and demystified information technology, it has also driven further change.

Speech Recognition

Speech and language (text) are presented separately at present, but two important directions that will become apparent in the next few years will be a real integration between them and the use of speech as a primary interface to the computer (eventually replacing the keyboard) (see Table 1).

The technology behind speech recognition and manipulation is very complex and relies on very fast, specialized integrated circuits. Some of the major bottlenecks in its inevitable advance will be:

- Really spontaneous, continuous speech recognition, including changes of speaker rate, emphasis, and accent.
- The use of telephone-quality voice input and applications requiring serious language processing. These both imply sophisticated levels of language understanding.

Poor environmental conditions (e.g., motor car, factory floor noise) will limit vocabulary and the interface presented to the user to, perhaps, a simple command language that maps onto preset commands or responses to prompts.

Table 1
Speech Recognition

Becoming Available by 2005	
Functions	Products
Speech recognition with 100K word vocabulary of carefully articulated speech without language interpretation	Better versions of today's dictation machines
Bulk voice data search and indexing for retrieval and manipulation	Limited spoken translation devices
Limited spoken language interpretation and dialog systems with 5K vocabulary	Handheld personal computers with voice input
Speech synthesis from text (no content understanding)	Speech input to computers; network-based speech services

Speed, Price, and Power

Civius, fortius, altius.
—Olympic motto

We now move on to some of the enduring trends in the world of computing and communication, the most important being:

- Ever-increasing processor power;
- Relative price of computer systems;
- Transmission cost and capacity;
- Storage price trends.

Ever-Increasing Processor Power

The computer has become ever more powerful. It has evolved over the last 30 years from an isolated curiosity, situated in a remote room and tended to by experts, to a ubiquitous and useful part of everyday life.

The two most important factors behind this dramatic advance in power and capability have been the increase in the sheer speed of processors and the growth in their memory capacity (see Figure 5).

To date, there has been a doubling of processor power approximately every 18 months or 2 years, a general trend that has come to be known as Moore's Law and which shows few signs of abating. Continuing developments in the silicon integrated circuit will ensure that this carries on well into the next century—device density has yet to hit the limits imposed by physics.

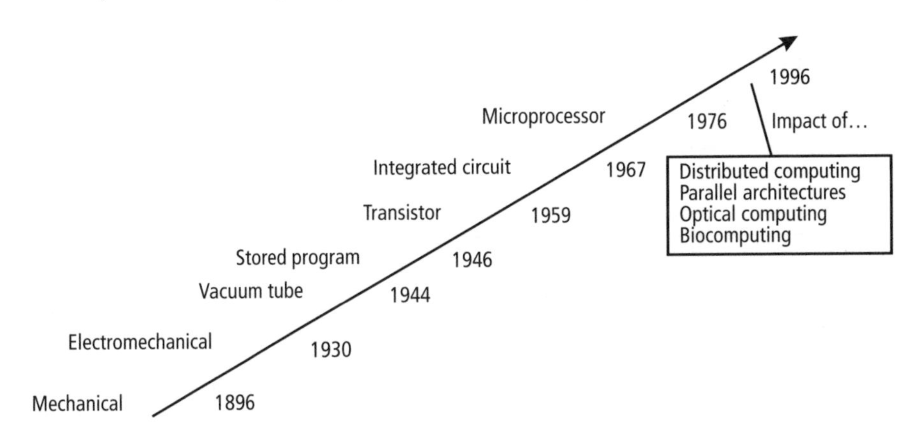

Figure 5 Ever-increasing processor power.

Of increasing importance will be the change to parallel and distributed architectures. The former will allow vast increases in the ability to solve problems without dramatic increases in the power of the processor chips. The latter provides the flexibility required to build and integrate more capable systems.

Relative Price of Computer Systems

The relative price of computer systems (i.e., processing unit per cost unit) has consistently fallen over the last 30 years (see Figure 6). This has been driven by two key factors:

- The underlying technology has become cheaper to produce, mostly due to continual improvements in manufacturing methods.
- Economies of scale have taken effect thanks to the sheer number of systems being produced (e.g., in the mid-1990s, more than 20 million PCs per year and 50,000 midrange systems per year).

As systems have become more standardized and open, competition between the vendors has also become extremely fierce. However, the overall costs of computing solutions have not fallen in equal measure, a consequence of complexity, the cost of integrating new and established systems, and the shift of cost toward the labor-intensive activity of software production. Nowhere are these effects more apparent than in telecommunications and information technology, where ever more money is required simply to keep pace.

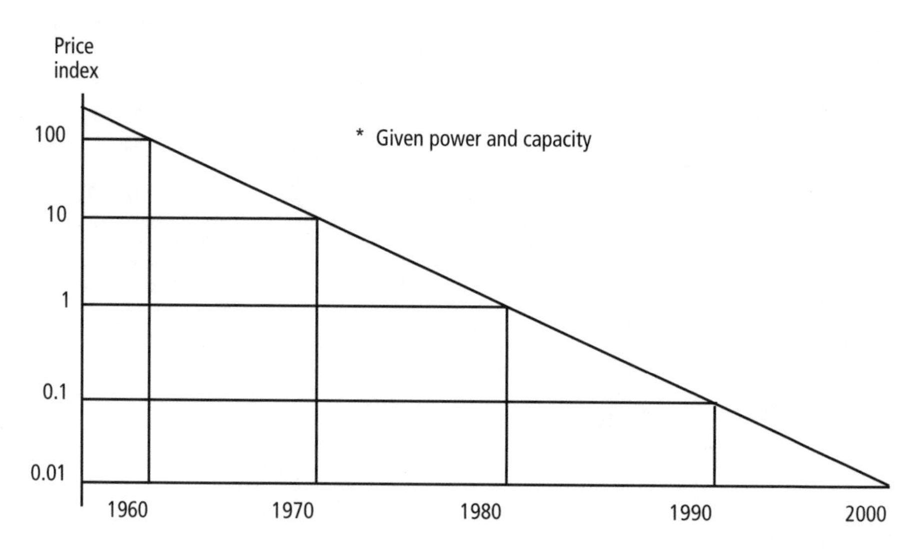

Figure 6 Relative price of computer systems.

Transmission Cost and Capacity

In general, information technology is increasingly concerned not with the processing of data (as it has been in the past), but in the movement of information from one place to another. The trend in this area has been consistent cost reduction coupled with greater capacity (see Figure 7).

This ever-increasing bandwidth (transmission capacity) at ever-decreasing cost means that more and more image and video information can be transmitted and applications that use them developed. Furthermore, the difference between local and remote operation is disappearing, and it is now more feasible to hold and manage data centrally and access data from the processing units wherever they may be physically located in a distributed environment. The boundaries of local- and wide-area networks are blurring as transmission technology makes total-area networking a possibility.

In general, there is a surplus of transmission capacity. By way of illustration, only 10% of the world's satellite capacity is used, and advances in compression technology make what is used more effective.

One problem that has arisen, though, is that of management. With increased distribution, network management is a growing challenge, and an assumption by many end users that bandwidth is free will drive a need for personal management (i.e., controlling the urge to send a huge video clip to a 500-recipient distribution list).

Storage Price Trends

Just as the price of processing is falling, so is the price of storage. Driven by the same base technology, it, too, is decreasing in cost by approximately a factor of 10 every 10 years (see Figure 8).

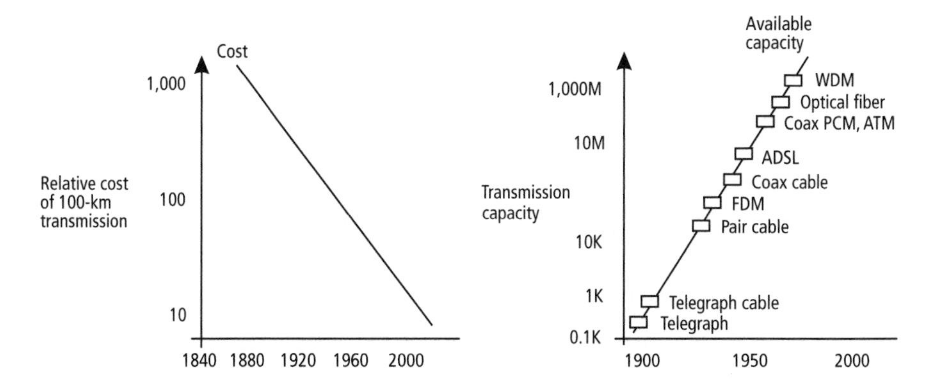

Figure 7 Transmission costs and capacity.

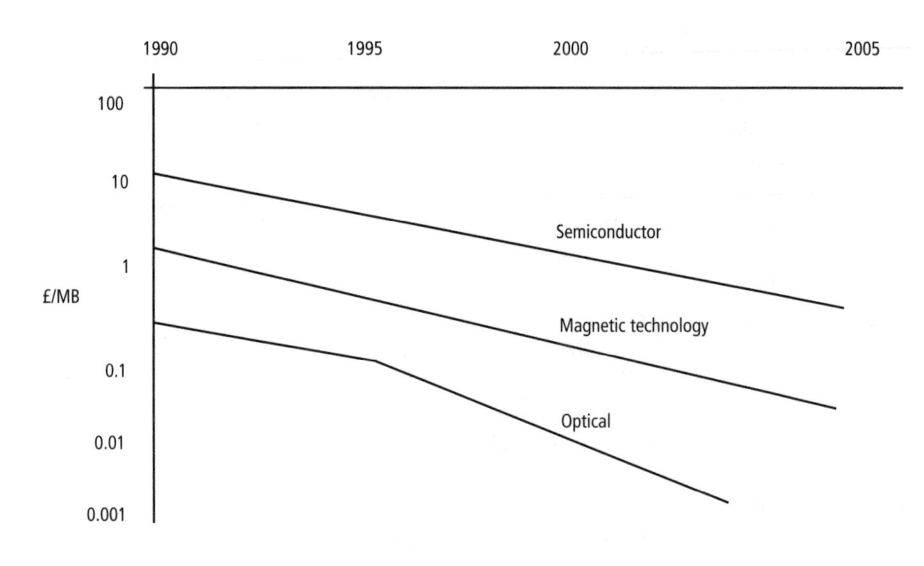

Figure 8 Storage price trends.

The trend is toward optical devices, which allow yet greater storage capacity, with data being packed more densely using three-dimensional storage techniques (a credit card–sized optical store can be used for 9 hours of video). Just over the horizon are holographic stores that promise to maintain the trends of recent years. The mature technologies such as magnetic core will continue to coexist in a speed- and capacity-based hierarchy (decreasing speed, increasing capacity) with the newer optical storage.

Fast online response systems will still use magnetic rather than optical storage. CD technology is fast approaching the stage where it matches hard disk access speeds.

Base Technology

The golden rule is that there are no golden rules.

—George Bernard Shaw

And so to some of the underlying drivers behind the current situation:

- Chips with everything;
- Overall technology characteristics;
- Memory trends by system;
- Battery developments.

Chips With Everything

All of computing and communication technology is founded on the integrated circuit or chip. As these devices get smaller, cheaper, and faster, so the systems they are used in get more impressive. Over the last 30 years, more and more has been crammed onto each chip (see Figure 9). Fabrication and production standards for silicon devices are now approaching the limits imposed by the basic physics of the materials used.

The net effect is that:

1. The storage capacity of chips will increase over the period to 1 Gb. This represents the storage capacity of 50,000 typed sheets (8.5 × 11 inches or A4) or 500 floppy disks.
2. The number of transistors per chip will continue to increase until the operational heat buildup proves to be a limitation. To reduce this, chips will work at lower voltages. We have already seen the standard chip voltage need fall from 15V in the 1970s down to 5V in the early 1990s, and 3V more recently.

With both memory and processing on a chip, many application functions will come as built-in hardware. Thus, some of the pressures to effect software control will be eased.

Overall Technology Characteristics

The figures cited in Table 2 simply substantiate the previous claims that both the number of gates on a chip and the number of bits that can be stored on a chip are increasing. The simple fact behind both of these advances is that the

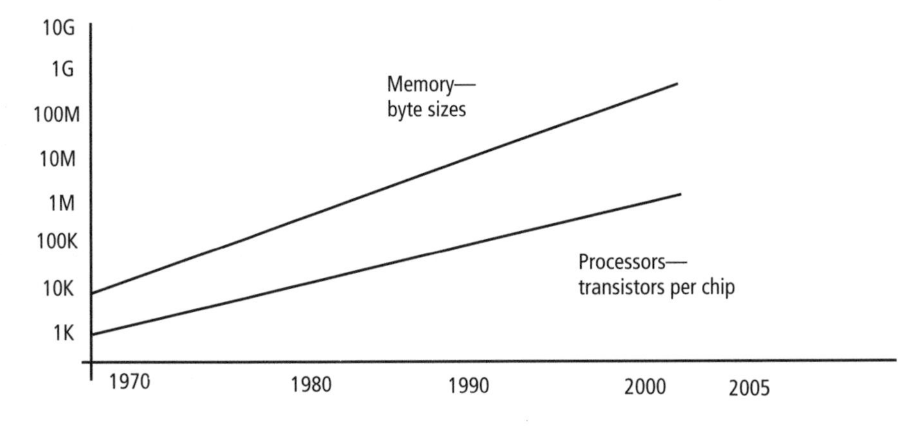

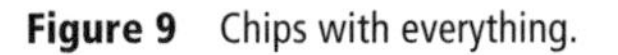
Figure 9 Chips with everything.

Table 2
Overall Technology Characteristics

Parameter	1992	1995	1998	2001	2004	2007
Feature size (microns)	0.50	0.35	0.25	0.18	0.12	0.10
Gates/chip	300K	800K	2M	5M	10M	20M
Bits/chip						
DRAM	16M	64M	256M	1G	4G	16G
SRAM	4M	16M	64M	256M	1G	4G
Performance (MHz)						
Off chip	60	100	175	250	350	500
On chip	120	200	350	500	700	1,000

resolution of the circuits is increasing (i.e., the number of circuits per square centimeter is increasing).

The manufacturing limits of the current process—photolithography—are fast approaching. Alternative means of producing chips have yet to be shown commercially viable. Optical and biocomputing have both been proposed, but have yet to deliver. Likewise, superconductors that operate close to room temperature have been demonstrated. All of these would have significant impact on the speed or functionality of the devices available to the system builder.

In the meantime, one additional factor that has had a large impact is the speed of the processors that are available. This is ever increasing and is likely to advance farther, from the typical 100 MHz today to a typical 500 MHz in 10 years' time.

Memory Trends by System

Although memory continues to increase dramatically, it nevertheless gets filled up very quickly by increasingly sophisticated operating systems and systems software (see Figure 10). There seems to be a perverse form of Parkinson's Law in operation that every PC user should be able to relate to—that the next generation of applications will invariably fill up the disk that you thought would last you forever.

In addition to this, more and more data are being manipulated directly in memory (particularly relational databases). Indeed, it is now feasible to load an

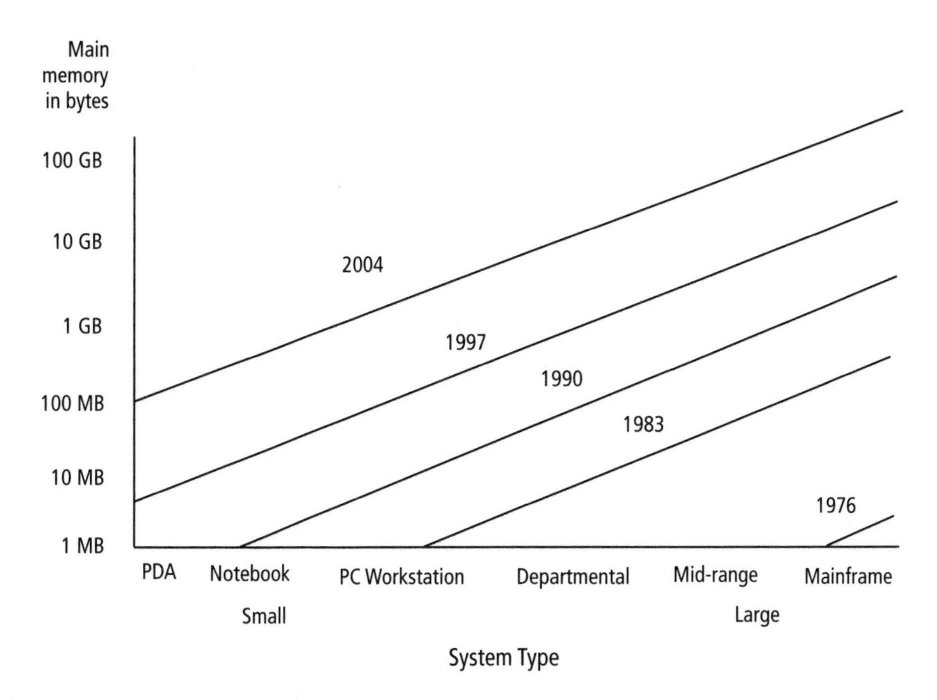

Figure 10 Memory trends by system.

entire database into memory prior to the application software accessing it. It seems that all memory on offer will be used somehow.

To cope with the increasing size and deployments of memory, 64-bit addressing architectures are now being deployed. These give a theoretical memory space of up to 16 million TB.

Battery Developments

The push for device portability is making battery technology a key factor. It is becoming increasingly important to have mobile devices that can operate for reasonable lengths of time, but that are not too cumbersome or heavy. Battery technology is therefore a very relevant part of modern information technology.

Significantly, the rate of improvement in this area lags behind that of other technologies very badly (see Table 3). This has broad consequences; for example, the constraining factor on many mobile devices is the capacity of batteries and their ability to be continually recharged.

In Table 3, it is noteworthy that although the capacity characteristics of Zn-air batteries are good, the number of cycles of charging/discharging (70)

Table 3
Battery Developments

Parameter	<1993 NiCd	1993–94 NiMH	1994 Zn-air	1996+ Lithium	1998+ SSPE
Working hrs/kg	45	52	157	93	200
Recharges	400	450	70	1,000	300
Charge, hours	1	1	10	2	5
Discharge, %/month	20	50	12	12	12
Cost (1=low)	2	4	3	5	5
Toxicity (1=low)	5	2	1	2	2
Safety (1=good)	1	2	1	4	3
Availability (1=good)	1	2	3	4	5

imposes a limitation that restricts the use of the batteries. Newer developments may help but resolution is not yet here.

Just as compression calms bandwidth demands, so there are ways of easing battery requirements. For instance, the choice of screen technology on a computer has a significant impact on power requirements.

Disposal remains a big problem, however. Recycling capability will prove increasingly important and will generate pressure for battery return and reuse.

Networks

The art of dialling has replaced the art of dialogue.
—Gitta Mehta

We now move on to the area of networks, once dominated by the telephone and voice communications, but now becoming the vital fabric for all manner of communications.

- The bandwidth explosion;
- The cost of calling keeps on falling;
- Local-area networks;

- Wireless data devices (U.S.);
- Networking issues 1;
- Networking issues 2.

The Bandwidth Explosion

As hinted at in several previous pages, there seems to be a never-ending demand for more and more network capacity. There are a number of reasons why this is so, the main ones being:

- Once simple transfers between computers have become much more complex and are also more frequently used. This is driven, at least to some extent, by the fact that the usefulness of networks increases as more people come on line. This is the basis of Metcalfe's law, which observes that the value of a network is vested in the amount of shared information.
- Rather than simple text files, much of the traffic now being generated is a mix of text, video, and voice. This has to be discriminated over the link and is inherently hungry for bandwidth (e.g., a picture can easily be 100K, a video clip 100 MB).
- Many users think little of attaching large files and sending them to long distribution lists. The ease with which anyone can generate enormous network traffic is noteworthy.

Given that demand has always risen to meet availability in the past (see Figure 11), there is every reason to expect the trend to continue. The message for network and system planners is clear—plan for growth!

The Cost of Calling Keeps On Falling

Just as the price of processing power and memory has plummeted, so has the cost of accessing public networks. It is not that long ago that the only way to get a high speed connection was to buy a leased circuit such as a T1. The introduction of the ISDN broke the 64-Kbps limit on "standard" local line connections and now the various flavors of digital subscriber line (DSL) technology promise to give the speed of a T1 at the price of an ISDN line.

This trend is illustrated in Figure 12, which gives the number of standard phone lines provided by T1, ISDN and DSL along with an approximate cost. The size of each bar indicates the cost per line—a clear sign that the cost of calling keeps on falling.

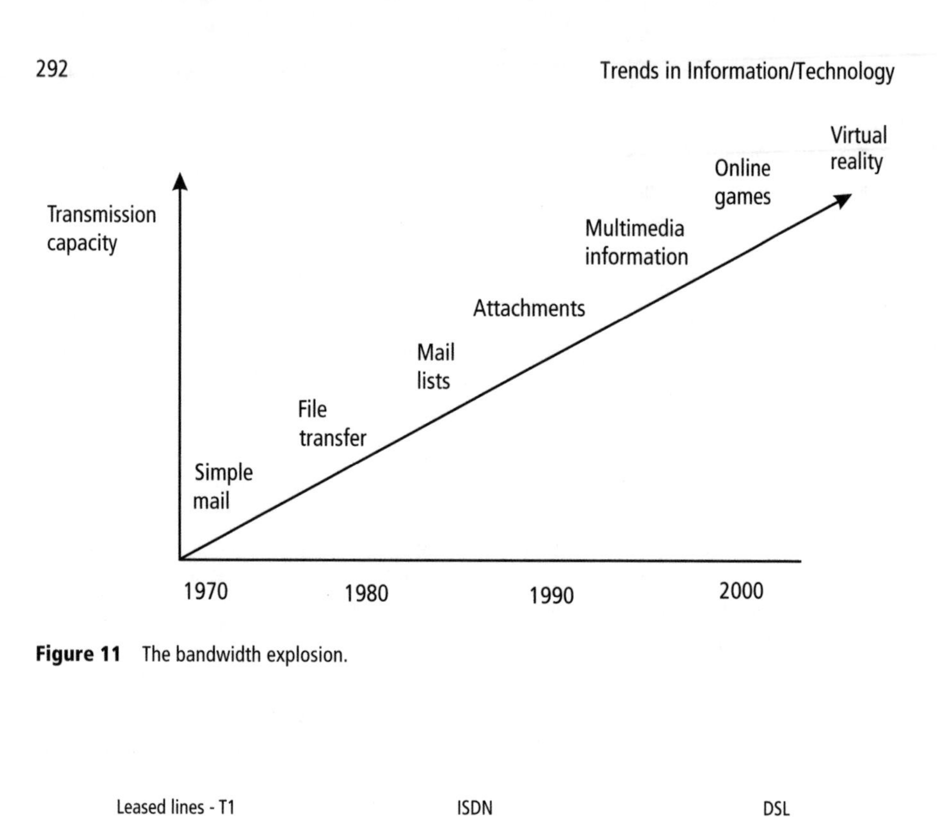

Figure 11 The bandwidth explosion.

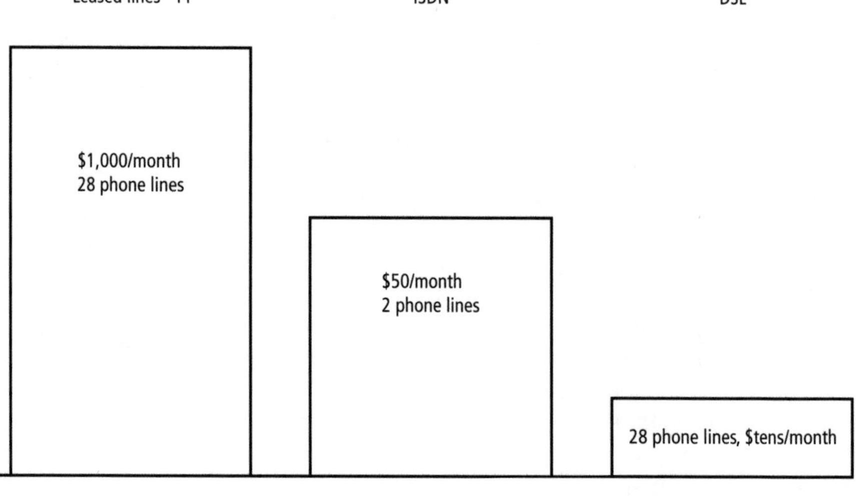

Figure 12 Access technology.

Local-Area Networks

For many people, the LAN has become a familiar part of the working environment (see Figure 13). Since the Ethernet was popularized in the early 1980s, it has been the predominant means of sharing information and resources in the office, the factory, and the laboratory. Despite the advent of higher speed LANs (e.g., FDDI-based), there have been few fundamental changes in local networking.

Asynchronous transfer mode (ATM) may change this. ATM will enable the transmission of voice, video, image, and data across the same communication link and therefore allow the true integration and management of these services.

Since ATM is becoming the important standard for wide-area networks as well as local ones, there will consequently be a coming together of LANs and WANs. This lays the foundation for globally distributed computing and is important for distributed organizations such as multinationals and banks.

The advance of ATM (which is predicted by Robert Metcalfe, the inventor of Ethernet) is likely to start at the local end with a public broadband data network emerging around the end of the century.

Wireless Data Devices—United States

The cellular phone has rapidly become a fact of public life. And it is now being developed to extend its usefulness through integration with pagers, computer, fax machines, and other devices.

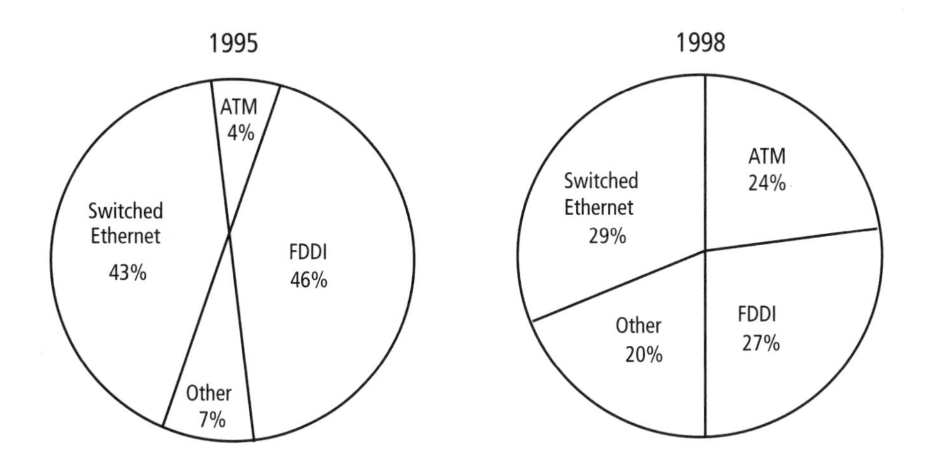

Figure 13 Local area networks.

Cellular technology is moving from analog to digital to make better use of bandwidth. This also opens the way for the wireless Internet, since many data services can make efficient use of digital cellular networks (see Table 4).

There will be an increasing use of wireless technology in vehicles, with a movement toward satellite transmission to give the necessary global coverage.

LANs in buildings will also use wireless technology, thereby overcoming restrictions in building structures and portability problems.

Networking Issues (1)

Networks underpin many of the changes that we are likely to see. Given that their path is dictated as much by politics and costs, it would be rash to make any sort of prediction based on technology alone. Table 5 represents a best guess at the way that the whole area is moving. Likely changes in the above aspects of networking will be an increase in the use of broadband, the adoption of wireless LANs, and a move to more lightweight data services such as frame relay.

Ethernet, with a transmission of speed of 10 Mbps, has been a standard for many years and will become available at up to 100 Mbps, further extending its use.

Copper wire, another well-tried technology, is also likely to remain a standard for local networks, and with the increasing sophistication of line equip-

Table 4
Wireless Data Devices—United States

Year	PDAs	Other Wireless Data Devices	Events
1994	100,000	500,000+	Radio modems and cellular phones on small cards are introduced for PDAs and PCs
1996	350,000	1,000,000+	Regional personal comms for data and voice are established
1998	750,000	2,000,000+	CPDP becomes a seamless national network, promoting transmission of voice and data on single devices
2000	1,500,000	3,000,000+	Handheld wireless data devices become mass market items

Table 5
Networking Issues (1)

Technical Issues	Current Issues	Future
Ethernet vs. token ring	Ethernet ahead	Likely to remain
Structured cabling vs. other options	Structured cabling dominant	No change
Smart hubs	Currently dominant	Likely to remain so
Wireless networks	Limited area of application	Expansion likely
Copper wire vs. fiber optics	Copper used when possible, fiber used for backbones	Change likely
X.25 vs. frame relay	X.25 ahead and likely to persist for some time	Frame relay will dominate
Broadband vs. narrowband	Narrowband dominant	Broadband usage greatly increases
Private vs. public networks	Private preferred	Virtual private networks seen as being more flexible and easier to manage

ment, will work at increasingly higher speeds, even over low-cost twisted-pair lines.

Networking Issues (2)

The important development that should be highlighted in Table 6 is the integration of separate services and information categories—such as data, image, voice, and video. The digitization and packetization of all these forms enables them to be carried over the same transmission lines. Thus, there is one multi-service network for all traffic, rather than a separate overlay for each different need.

In the world of open systems, the formal standards issues by the OSI have been slow to be adopted because of the universal popularity and availability of the Internet, TCP/IP, and other de facto standards. However, formal standards are likely to gain slowly, particularly in WANs, since this is where secure international consensus is important. Part of this is already evident in the area of network management, where the scale of evolving networks demands a con-

Table 6
Networking Issues (2)

Technical Issues	Current Issues	Future
ISDN usage	Slow introduction	Will accelerate
ATM	Limited availability	Will become dominant
LAN and WAN	Separate systems	Likely to merge (via ATM) to give total-area networks
OSI vs. TCP/IP	TCP/IP leading	Coexistence likely, convergence expected
Voice/data/video	Carried separately	Will combine
Network management database	Usage increasing	Will use object databases rather than RDBMs
Security of transactions	Internet style freedom from control	Greater access control and content policing

sistent approach. Here, object databases will extend to store maps, pictures, and diagnostic tools, as well as the usual status, alarm, and error information.

Software and Computing

There are only two commodities that will count in the future. One is oil and the other is software. And there are alternatives to oil.
—Bruce Bond

With so much advanced equipment on offer, it becomes more and more difficult to use it to good effect. Software has been the primary means of controlling and applying technology. The issues here are:

- The rise and rise of software;
- Application development;
- Who cuts the code;
- The progress of computing;
- Convergence;
- Divergence.

The Rise and Rise of Software

System costs have become more and more dominated by the cost of software production over the last 20 years (see Figure 14). Most of the problems in the IT industry over the last 10 years (notably the spectacular failures) can be traced to software problems.

This concern has been the main driver for improvements in software quality (e.g., ISO9001, TickIT, and SPICE standards) and the wider deployment of methods oriented toward "standard" software components (e.g., object orientation).

Tables 7 and 8 illustrate the nature of the software problem.

With hardware costs already approaching the marginal and software development starting to come under control, the next major system issue is likely to be the integration of disparate components.

Application Development

The first diagram in Figure 15 illustrates the ever-increasing help that has been made available to the software developer. Improvements in programming languages and operating systems have enabled the software engineer to produce a

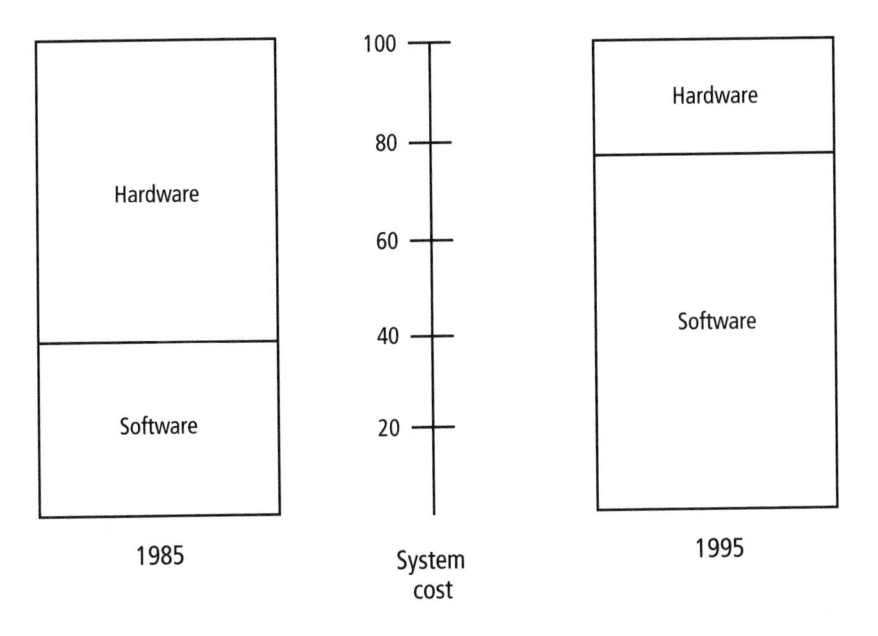

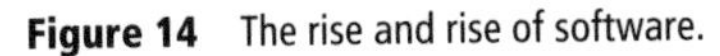
Figure 14 The rise and rise of software.

Table 7
The Growth of Software Dependence

Application	1985 Lines of Code	1995 Lines of Code
Car	Less than 5K	Over 30K
Telephone exchange	About 1M	From 5M to 25M
Aircraft	About 400K	About 5M
Television	Less than 10K	Up to 500K
Database	About 500K—accessing 100 MB data	About 4M—accessing up to 5 TB data

Table 8
The Software Delivery Problem

Project Size (Million Lines)	Average Delay	Cancellation Probability
0.1	2 months	5%
1	1 year	14%
2	1.4 years	18%
5	1.8 years	25%
10	2 years	36%
20	2+ years	40+%

better quality product and produce it faster. At the same time, the cost of the support tools has contributed significantly to overall costs, and in some cases, the production technology has caused more problems than the original task.

With the demand for software control spiraling, the overall picture over the last 30 years—shown in the second diagram of Figure 15—is of software costing too much and often not doing what was wanted.

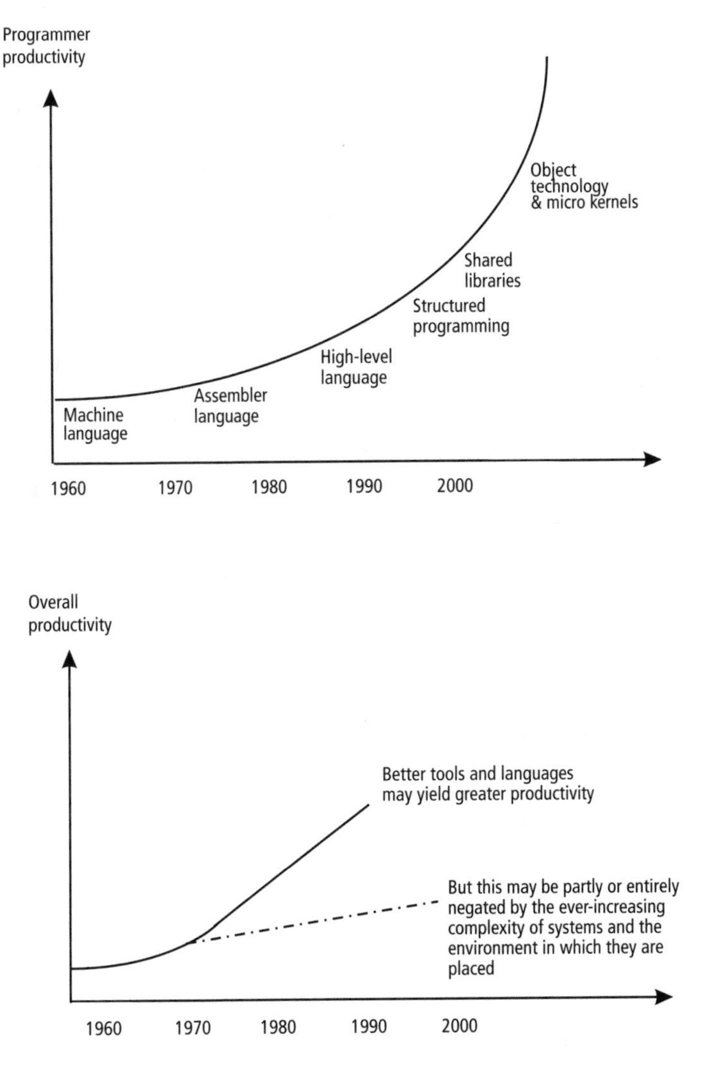

Figure 15 Application development.

To reduce the cost of developing new applications, it has become necessary to try and standardize software modules and to reuse them where possible. This promises a fundamental growth in productivity.

New applications will therefore not be written from scratch, but rather assembled from standard modules to a consistent overall plan. The modules may originate from different code builders, but due to the emerging standards they should all work and interwork with each other. The standards that will enable this to happen will be broad, covering both architecture (structure of distributed computing environments, remote procedure calls, interface definition language) and components (definitions of standard objects, object brokers, code libraries).

The drive is to build open flexible systems that can be built with products from a range of vendors and are easy to maintain and upgrade.

Who Cuts the Code?

The production of software has, in the past, been the preserve of the talented specialist. Many of the early programming languages and operating systems required considerable skill of their users.

Things have moved on, though. The pressure of more program control being required than people available to do it, along with the spiraling costs of software production, has caused a significant shift. Many of the languages and tools that can be used to generate useful code are now accessible to the nonspecialist. The personal computer, object-oriented methods, visual languages, application builders, and the like have allowed a much broader range of people to produce software (see Figure 16). The main groups who will write the programs will be:

1. End-user programmers who program to accomplish their job, often without realizing that they are producing software.
2. Professionals in software companies—the traditional expert programmers who work for software vendors such as Microsoft, ICL, Oracle, Admiral, FI, CAP, and Logica.
3. Professionals in product companies who produce code that is embedded in some product and sold to a customer, but not primarily as software (e.g., IBM, Digital).
4. IT departments are programmers whose code is used by their employer for internal purposes—for example, customer handling, support systems, value-added functions. Increasingly, companies are outsourcing this type of work to programmers in 2 and 3 above.

The Progress of Computing

The computer was once a specialized machine that required experts' attention to work and a controlled environment in which to operate. Nowadays, it is a familiar and ubiquitous tool that has permeated virtually every aspect of our business and personal lives.

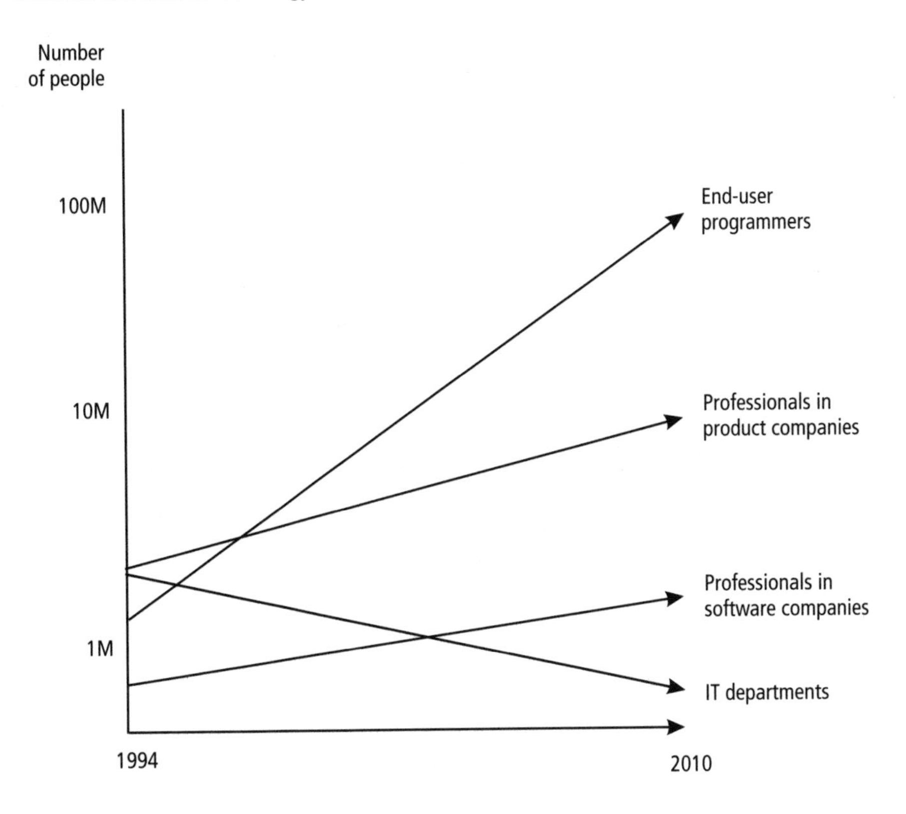

Figure 16 Who cuts the code?

Some of the more notable aspects of the progress of the computer are given in Table 9. This is something of an abstraction of a more complex and convoluted reality. Nonetheless, one thing that should be very clear is that there is a general path, down which computers are likely to move further—easier to use, more powerful, and more mobile.

Other parts of this section elaborate on specific aspects of this. The general trend, though, is toward dependence, not only on the computer, but also on its continued development.

Convergence

One of the long-term goals of the communications and computing industries has been to have systems that are truly open (see Figure 17). This means that a computer from one vendor connects to a network device from another and to a computer from another still. Standards institutions (e.g., ISO) and industry

Table 9
The Progress of Computing

Computing Era	Batch (60s)	Bureaus (70s)	Desktop (80s)	Network (90s)
Technology	MSI	LSI	VLSI	VLSI/DSP
Location	Computer room	Terminal room	Office	Mobile
Users	Experts	Specialists	Professionals	Anyone
User position	Subservient	Dependent	Independent	Free
Data	Machine oriented	Text	Graphical	Multimedia
Purpose	Calculation	Compute	Present	Inform
Interaction	Submit jobs	Interact (slowly)	Click and point	Consult and tell
Connection	Peripherals	User terminals	Shared resources	Personal devices
Applications	Custom	Standard	Generic	Component
Languages	Assembly level	Program level	Application level	Object oriented
Operating system	Single threaded	Timeshare	Multitasking	Client/server

consortia (e.g., X/Open) have invested a lot of time defining workable standards for open systems. This is a trend that will continue until networked systems can be plugged together from a set of component parts.

The lead in building a common model for communicating systems has come from the international standards body, the ISO. Its seven-layer model is widely used as a reference for different services and protocols. In addition, a number of initiatives from the computing community that deal with interoperation (as opposed to interconnection) are now coming to fruition.

It can be expected that standards, based on the likes of the DCE, will form the basic shape of future information systems.

Divergence

There will always be a need to differentiate between one product and another. So, despite there being a generally agreed-upon framework into which components fit, there will be huge diversity among those components (see Figure 18). Furthermore, the choice of available components, types of applications, and possible configurations means that virtually every real system will have some degree of uniqueness.

The drive to standardize in communications and computing supply has led not to uniformity of systems, but rather to choice of implementation. The

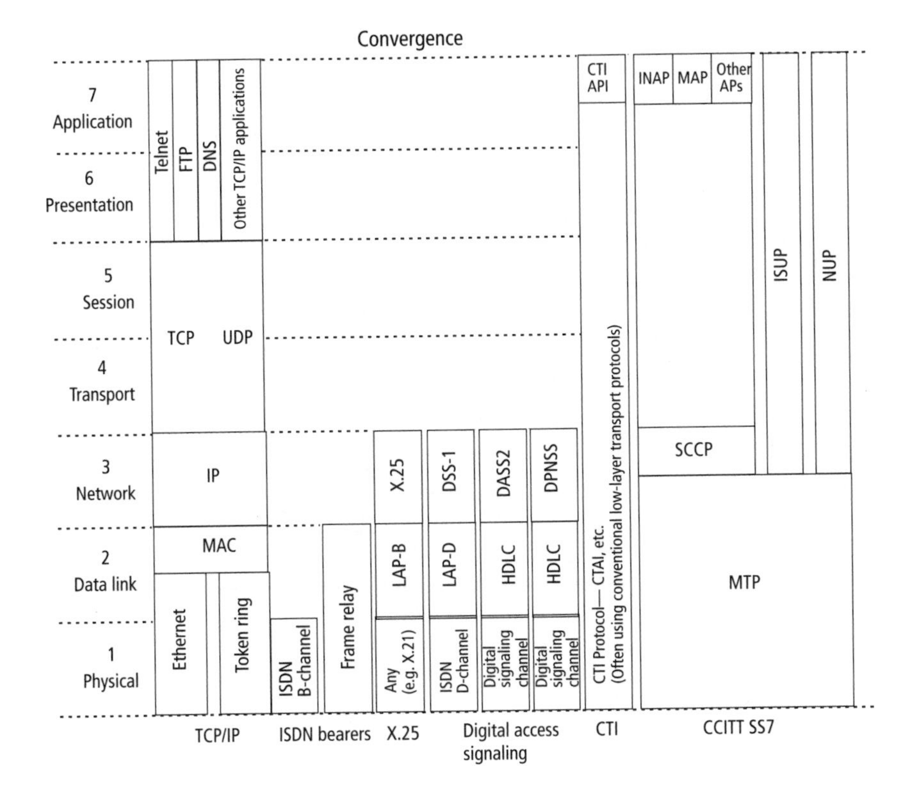

Figure 17 Convergence.

adoption of open systems has not taken away the need for careful systems planning and analysis—it has moved it to a more abstract level. So the move toward "mass customization" (an ideal envisaged by Joseph Pine in the early 1980s) looks like one that will continue.

The Growing Global Internet

If you're not on the Net, you're not in the know
—*Fortune* Magazine

Like it or not, you ignore the Internet at your peril. It is a truly global phenomena that impacts us all. Important issues are:

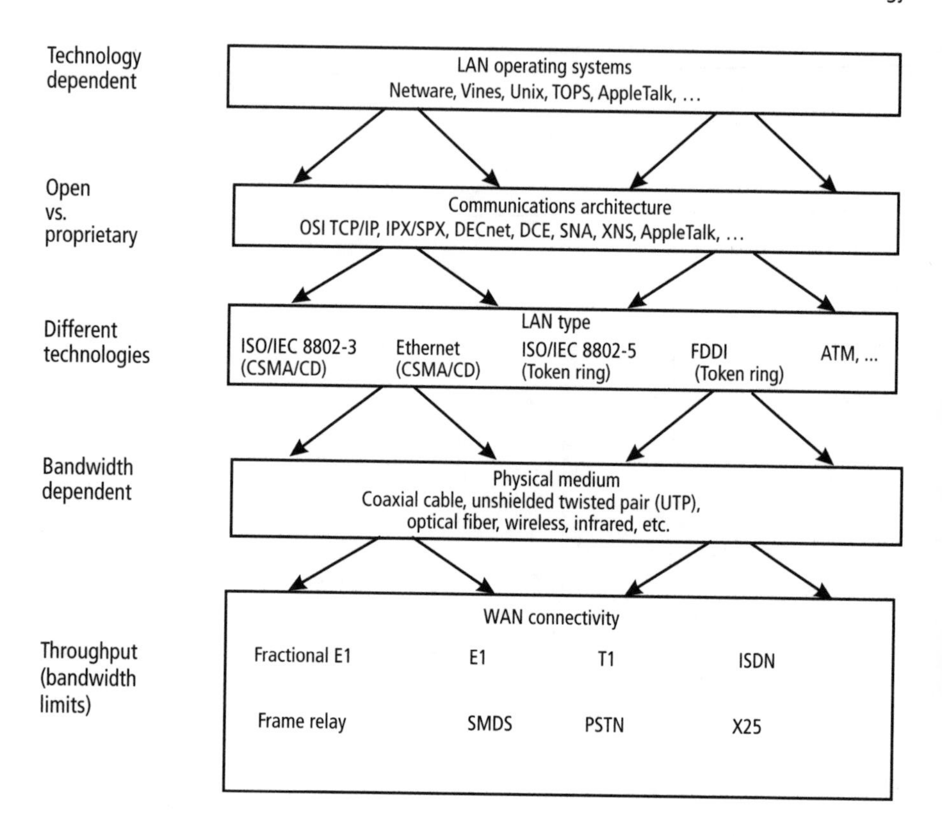

Figure 18 Divergence.

- The rise and rise of the Internet;
- Intranets and extranets;
- Infosurge;
- The information superhighway.

The Rise and Rise of the Internet

It is somewhat difficult to say exactly how many people are connected to the Internet—estimates vary according to the assumptions made, the person making the estimate, and the time of day. One thing is clear, though—that there is a huge and growing community of users connected to a worldwide information resource.

The significance of such a large group of people who can share ideas and information at will is not yet clear. There is a simple message that can be taken from Figure 19, and this is that it is now essential to be connected. Failure to do so leaves both the individual and the organization cut off from much of the mainstream of technology.

Extrapolating the growth in Internet connections over the last few years, it can be predicted that the whole world will be online by the year 2012. Unlikely as this is to pass, there are few signs of abatement in Internet growth and the trends of the past years can be expected to persist for the immediate future.

Intranets and Extranets

Businesses are becoming more and more decentralized. As they do so, in the process acquiring partners, collaborators, and joint ventures, they are starting

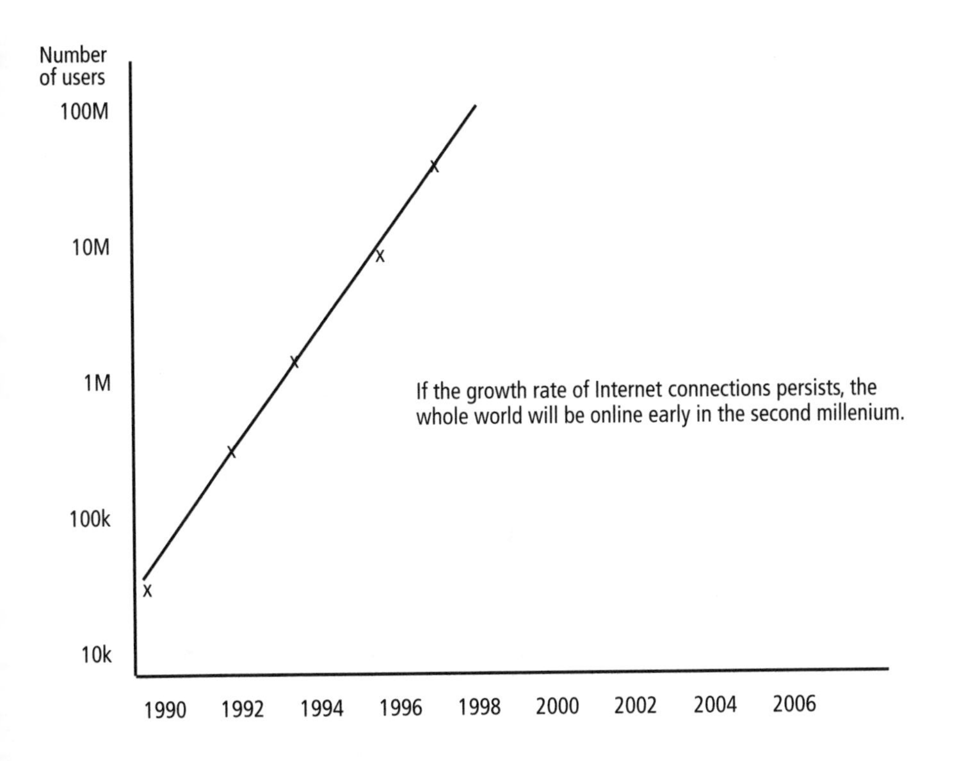

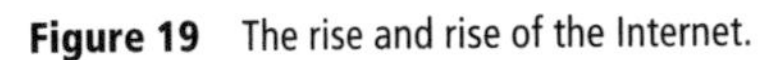
Figure 19 The rise and rise of the Internet.

to look more like federations of autonomous enterprises rather than a traditional organization. This is illustrated in Figure 20.

In order to cope with this, the networks upon which they depend to carry and process information will have to be capable of growing, shrinking, or changing shape as the situation demands.

To achieve this, these network will have to adopt an architecture that separates out the ephemeral from the more permanent (see Figure 21). For instance, the servers in the network may be tiered into corporate (which holds central company data), business (which effects the current business processes), and object (which serves the working applications and data). There should be flexible partitioning of software across desktop, business, and corporate servers. The user should have transparent access to data without needing to know if they are stored locally or remotely.

Increasingly, this capability is provided through intranets—corporate networks based on Internet technology. The way in which these have gone from nowhere to a billion dollar business in just a few years is shown in Figure 22, and the established trend is predicted to persist for several years to come. Increasingly, companies are linking their intranets with those of their partners to produce secure information domains known as extranets.

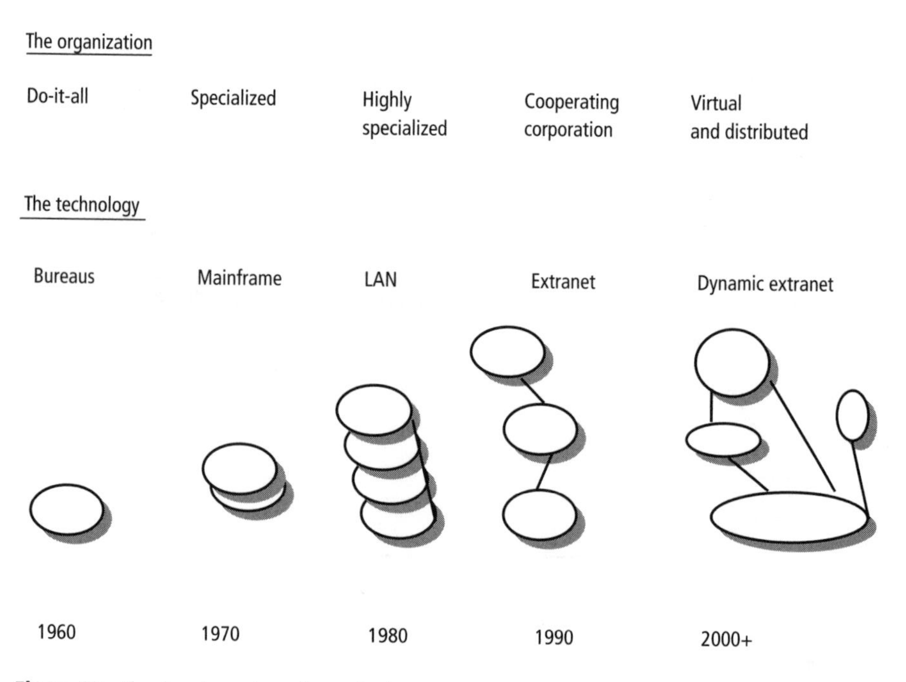

Figure 20 The changing nature of organizations.

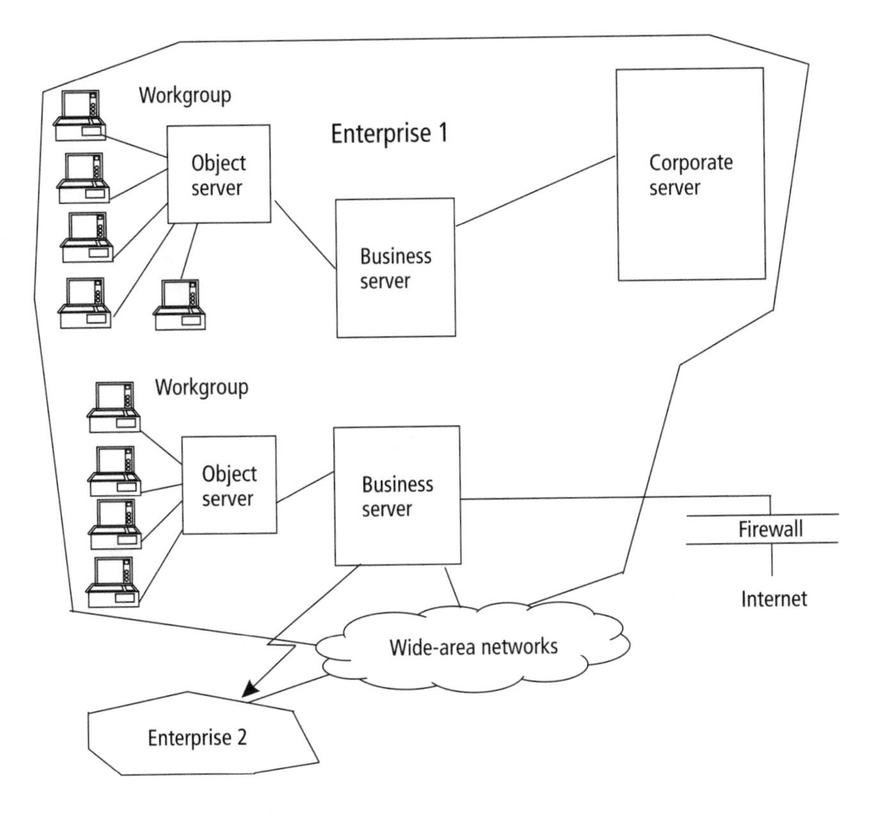

Figure 21 IT for the enterprise.

Infosurge

With such an expansion in the Internet and the same technology being increasingly used in the workplace, there has been an explosion of online information. The growth in World Wide Web sites, shown in Figure 23, has provided a huge, global source of information. Furthermore, the ever growing proportion of addresses with the .com extension indicates a trend towards commercial use. No longer is the Internet the preserve of academics, nor is it constrained to national boundaries.

In terms of use, there have been significant developments with the Internet, as shown in Figure 24. In its early days, it was used, in the main, by experts to transport files between computers. This is reflected in the high proportion of the ftp application that was used. As the Internet has become more popular, the trend has been to use Web transactions (i.e., browsing and other activities that can be carried out from within a browser) to take over from ftp, with mail and other applications remaining fairly steady in their use.

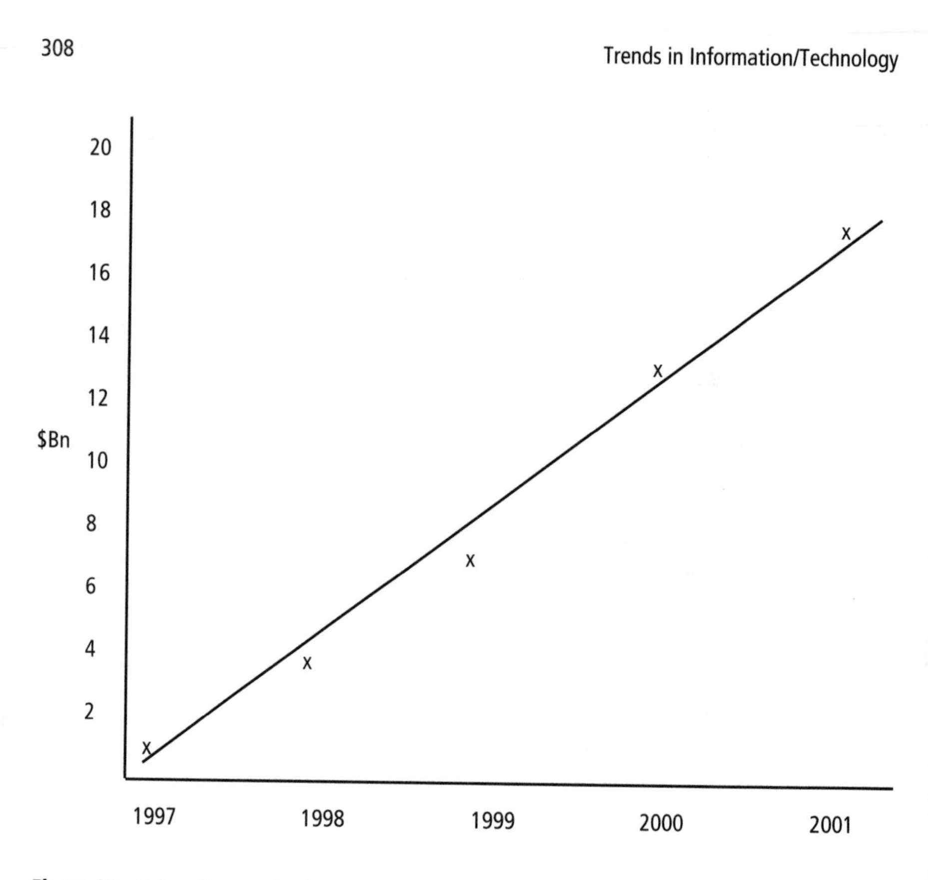

Figure 22 Value of Internet business.

One of the inevitable consequences of so much information, so readily available is that people will have to be increasingly discriminating in what they read. Figure 25 illustrates the onus placed on the information user. With information a great consumer of the user attention, mechanisms for summarizing, structuring, and selecting will have to be developed.

The Information Superhighway

The information superhighway is rapidly becoming one of the most used, and abused, terms in the English language. Everyone has a different idea of what it really is or will be. Figure 26 indicates some of the important features and components of the information superhighway. The general picture is of an array of services provided over a variety of networks to a diverse set of connected devices. So there is not one uniform superhighway, but rather a range of products and capabilities that will be available for the purpose of information sharing and processing.

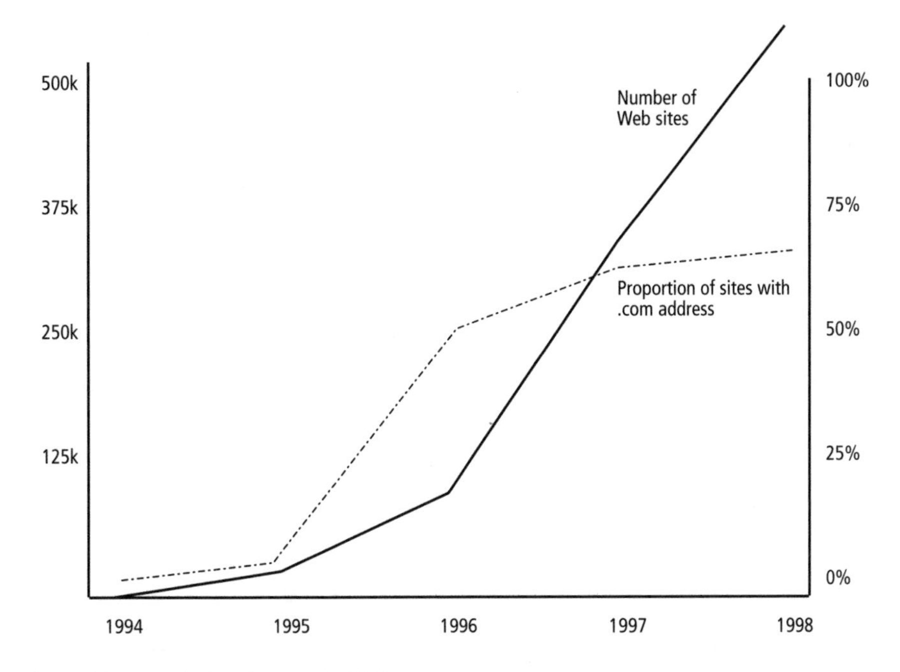

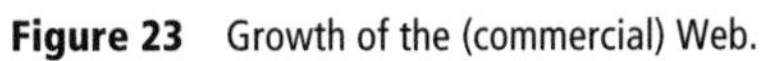
Figure 23 Growth of the (commercial) Web.

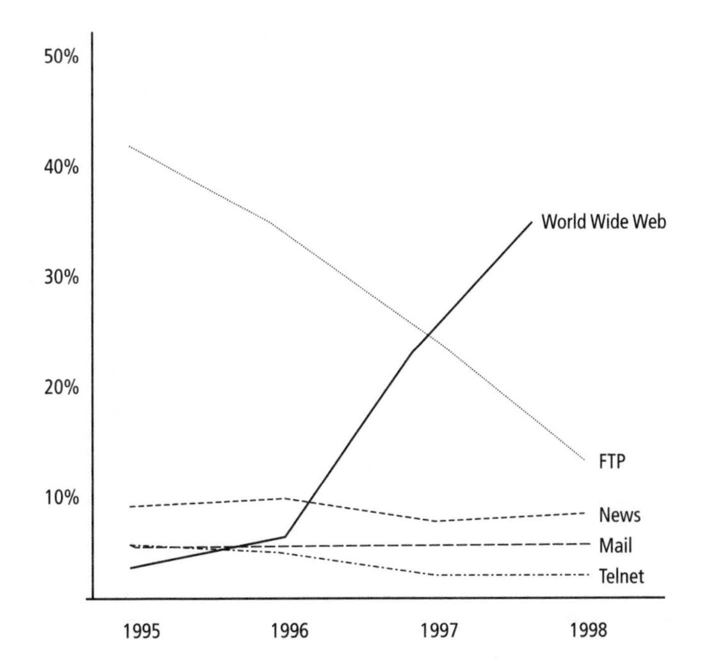

Figure 24 Changing usage patterns on the Web.

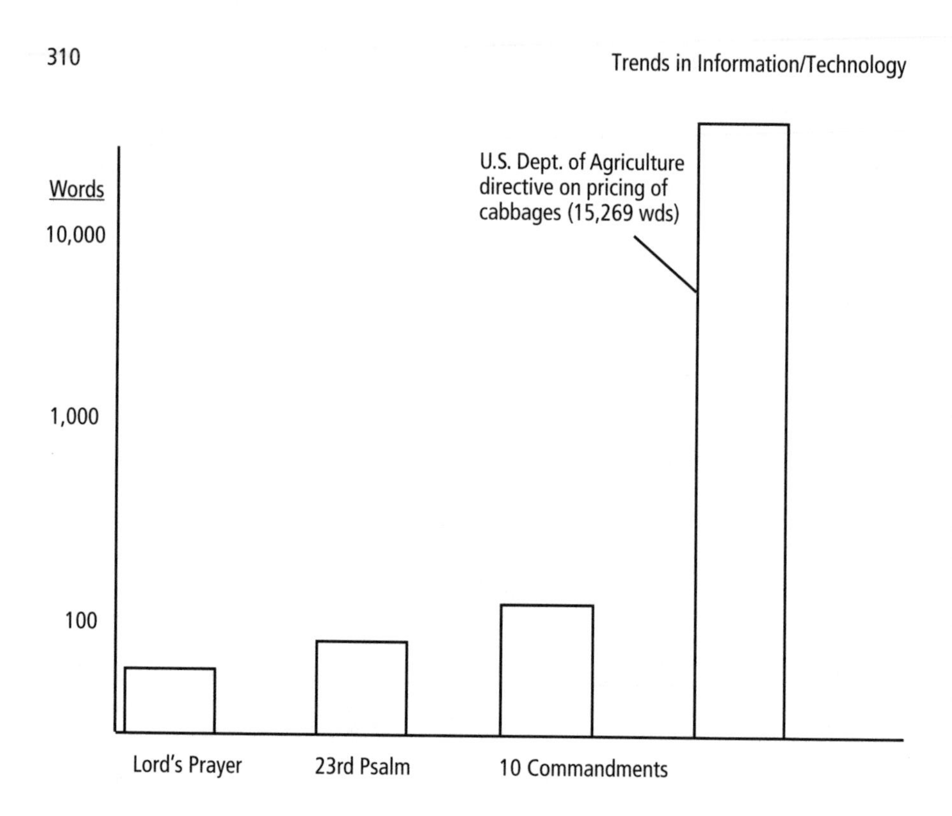

Figure 25 The growth in complexity.

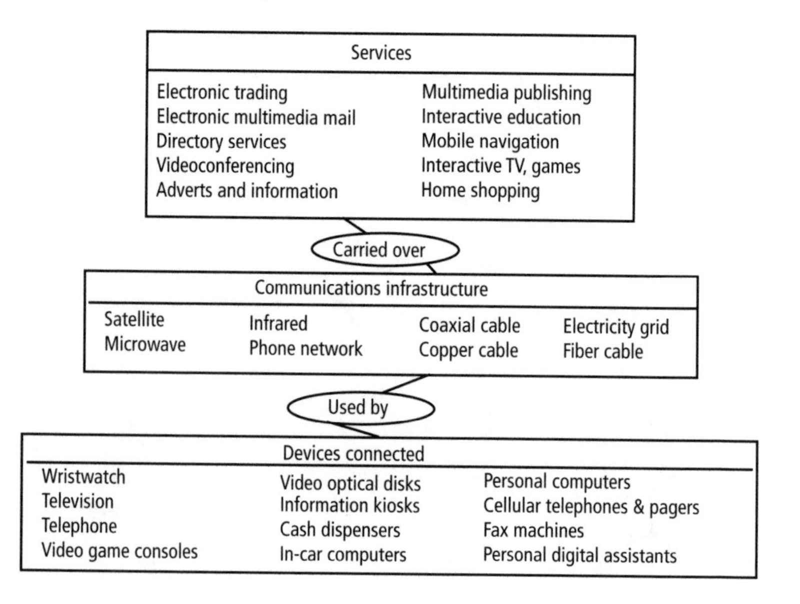

Figure 26 A technology hierarchy.

This will not happen in a flash of innovation or investment, but will gradually develop through the expansion of the existing communications infrastructure and services. The communications infrastructure that forms the basis of the superhighway includes high capacity networks for academia (such as Mbone, DANTE, and SuperJANET), the Internet (already a worldwide network linking research, academia, business users, and others), government networks, and various commercial networks (such as the ISDN).

To finish with, Table 10 gives some general figures on the level and distribution of telecommunications and computing resources across the world. Although taken as a snapshot in time (1996 to be specific), the contents of the table do give some context for some of the trends given in the previous section.

The top half of Table 10 gives the number of people in every hundred who have access to their own telephone line, fax machine, and cable TV link, and who are offered a digital network with international dialing. The second half of the table lists straightforward numbers of Internet hosts (e.g., unique IP addresses), commercial satellite transponders, payphones, and private circuits.

Given the diversity across any one of the areas in the above table, it is not possible to draw any particularly precise observations. There are several possible

Table 10
A Snapshot of World Telecommunications

Parameter	North America	South America	Western Europe	Africa and Mid-East	Eastern Europe	Asia	Oceania
Lines	62%	8.5%	44%	3%	16%	4.5%	39%
Fax	2.8%	0.17%	1.4%	0.04%	0.1%	0.02%	1.92%
Cable TV	63%	0.2%	8%	0.15%	1%	2.3%	0.01%
Mobile	10.5%	0.5%	4%	0.1%	0.1%	0.4%	11.6%
Digital network	67%	52%	65%	68%	17%	66%	62%
International dialing	100%	92%	100%	74%	76%	82%	100%
Internet hosts	5.5M	50K	1.8M	63K	67K	270K	290K
Satellites	720	85	223	107	29	465	31
Payphones	2M	0.76M	1.4M	0.16M	0.4M	1.8M	0.05M
Private circuits	15M	0.1M	3.8M	0.12M	0.05M	1.3M	1.2M

scenarios for the practical deployment of the information superhighway. Some of the options for the user end are illustrated in Figure 27. No one yet knows how this potentially huge market will develop but it is not unreasonable to suggest that the television will become the universal communications device for the family. And, in a digital world, the variety of applications that can be presented through that device will be limited only by the imagination and the availability of decent software.

Whatever path is taken, there have always been local preferences, and it is likely that variations across the globe will persist. Some of these can be attributed to economic or geographic differences. Others are more a matter of taste or preference. Together with the technical information presented earlier, this is a useful additional pointer to understanding the world of computers, communications, and the Internet.

Mobile phones become portable communication devices thanks to a combination of applications that can be delivered via mobile code and bandwidth from plentiful satellite capacity.

Personal computers evolve to become network computers. Smart cards enable secure access and mobile code users to get all the programs they need, as they want them, over a network rather than having to store them on a personal computer.

The television becomes the universal communications device for the family. Set-top boxes give the established TV the ability to access the Internet and the browser becomes the universally accepted interface to online information.

Figure 27 Variety of network access devices.

Organization

One of the hottest frontiers in technology is for systems that will enable people to cooperate across national and economic frontiers.
—*Fortune* Magazine, 1988

Finally, there are some general trends and developments that just cannot be ignored:

- The changing organization;
- Patterns of work;
- Technology for distribution.

The Changing Organization

Technology and the people it serves cannot be neatly divided. There has been a trend for organizations to become more distributed and less rigid than they have been in the past (see Figure 28). Part of this may be attributable to the technology that makes it feasible. On the other hand, it may be the change in organizational structure that has driven the technology to support it. Either way, there are some continuing trends.

The seven organizational shifts in the information age are:

	From	*To*
1.	Rigid	Flexible
2.	Preallocated responsibilities	Collective responsibility
3.	Top-down decisions	Decisions focused around key information holders
4.	Leadership vested in rank	Leadership dependent on situation
5.	High cost of entry and exit	Low cost of entry and exit
6.	Run to defined procedures	Customized procedures
7.	Face to face predominates	Electronic communication predominates

Patterns of Employment

The demise of secure employment for life has been predicted for some time. Commentators have talked of a "portfolio lifestyle," where individuals combine

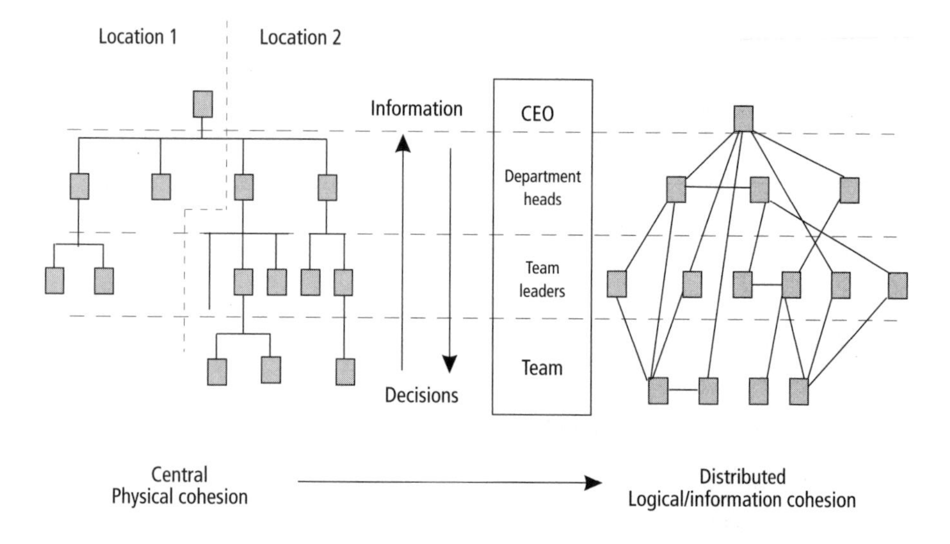

Figure 28 The changing organization.

several activities (leisure being one of them) as best fits their capabilities and needs. This is one idea that is explored in Charles Handy's excellent book, *The Empty Raincoat.*

Signs of this predicted trend can be discerned in the increasing number of home workers, work sharers, and outsourcing contracts. The general shift is toward the individual, away from the company or organization. Figure 29 shows some of the patterns of employment expected over the next five years.

Part of the drive for more fluid working practices comes from a desire for flexibility. Organizations want to be able to respond more quickly to changing needs, so they want fewer fixed structures. The other part of the drive is technology that makes it possible to work wherever a person wants. Given that both the individual and the organization have something to gain from the trends shown above, they are likely to continue.

Technology for Distribution

The Internet model of federated subnetworks that share an addressing scheme and communication protocols has grown enormously over the last 10 years.

The traditional public switched network has evolved much more slowly, by adding more intelligence and features to an already well-engineered, centrally controlled and administered system.

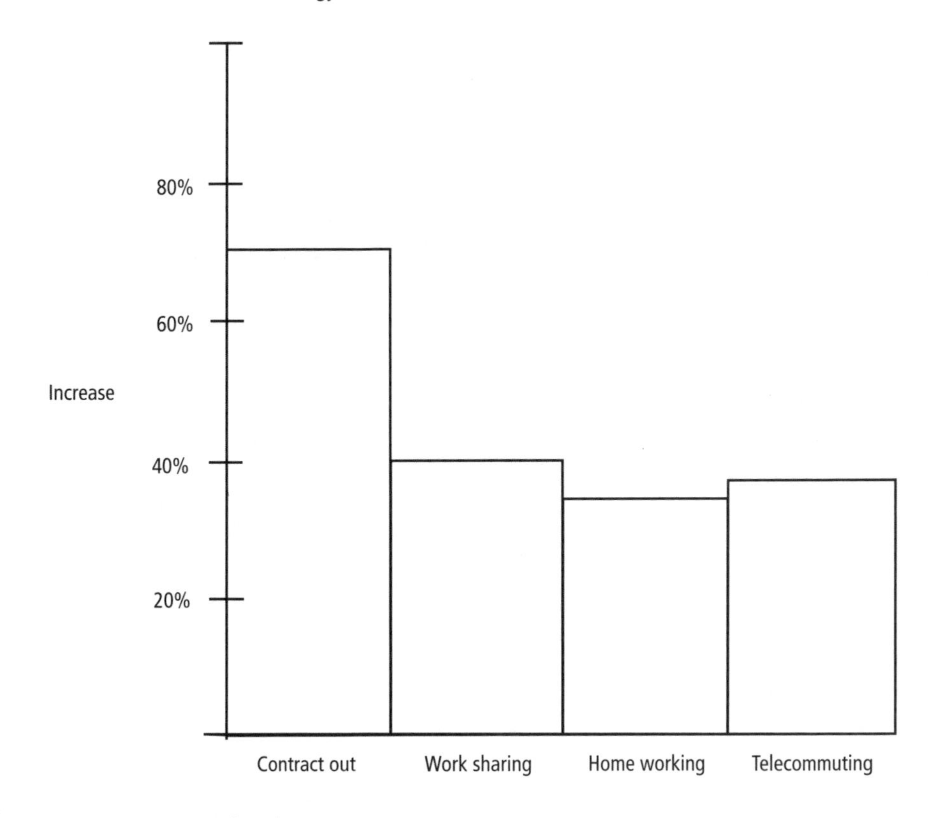

Figure 29 Patterns of employment.

The contrast of scalability and assured quality of service (the connection-oriented view) and fast and flexible (the connectionless view) provides the designers of networks with a choice of method for distributing systems.

Each poses its own challenges. For instance, each has very different characteristics when it comes to managing the network—the required information can be strictly controlled in the first of the above, but can be arbitrarily distributed in the second (see Figure 30).

Seven Challenges

To close this section, we draw together some of the general technical challenges, the often mundane details that will probably limit the practical exploitation of evolving technology.

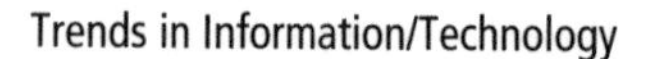

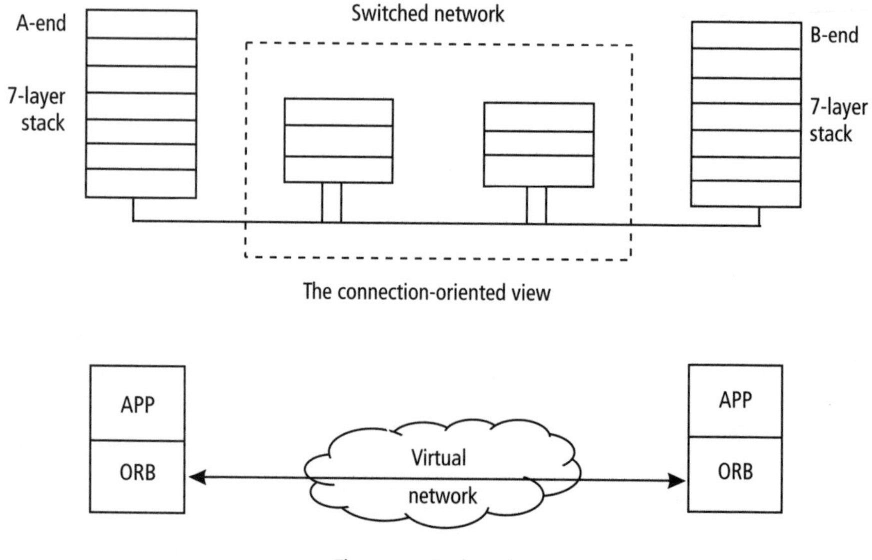

Figure 30 Technology for distribution.

1. Power

Mobile products are increasingly called for to support distributed organizations. Batteries are not yet good enough.

2. Weight

Present packaging technology is too heavy for convenient portable products. A reduction in chip operating voltages seems to be the best route to get round the main source of the weight—batteries.

3. Cooling

Cooling is a problem with current packaging. Again, the solution is to minimize bulky heat sinks, to reduce operating voltages to 1V or below, and to eliminate cooling fans.

4. Cabling

Cabling between computing and communication equipment presents a potential barrier to flexibility. Radio, audio, infrared, or optical links could all eliminate the problem and allow the user access to information wherever it may be. Depending on the required range, IRDA (Infrared Data Association, a convention for line-of-sight links) offers one solution, wireless LANs another, and microwave and GSM networks still others.

5. Complexity

System complexity is a major log jam, particularly in the software needed to drive large systems. The most likely solution here will come not from better production techniques, but rather from more standardization of system components and structure. The extent to which open systems are adopted and supported will play a major role in dealing with complexity.

6. Security

The increasing degree of interworking will focus attention on the problems of security in computer systems. It is easy enough to provide security when transmitting information. The real problem lies with the ability to protect against subversion. The dramatic increase of the online community (mostly via the Internet) will force attention on both computer and network security.

7. Efficiency

Current software is inefficient, both in its use of memory and of processing power. The widely used Windows operating system is an order of magnitude more profligate in its use of memory and processing than its predecessor and, arguably, has only been redeemed in the eyes of its users by headlong advances in hardware.

While the hardware continues to advance at the current rate, there is little economic pressure to improve the software situation. In fact, there are cases where hardware options have been preferred to software primarily on grounds of efficiency (e.g., ATM for high-speed networks).

Two factors may become apparent in the foreseeable future. Hardware improvement will appear to diminish in relative terms while software design will be forced to become performance-oriented from the earliest stages.

Bibliography

Given the breadth of computing and communications technology, it would be virtually impossible to give a complete reading list for anyone who needs to look a little deeper into specific topics. There are, however, a few texts that have proved very useful for building a broad appreciation of the subject area. These are not necessarily the definitive texts, found on every university reading list. Rather, they are the books that present a broad view of some aspect of technology and its application and relevance. So, to delve a little deeper into any of the areas covered in this chapter, the following are recommended:

The User and the Marketplace

Drucker, P., *Post Capitalist Society*, Butterworth Heinemann, 1993.

Naisbitt, J., *Global Paradox*, Nicholas Brealey Publishing, 1994.

Porter, M. E., *Competition in Global Industries*, Harvard Business School Press, 1986.

Reich, R., *The Work of Nations*, Simon & Schuster, 1991.

Speed, Price, and Power

Davies, D., C. Hilsum, and A. Rudge, *Communications After AD 2000*, Chapman and Hall, 1993.

"The Computer in the 21st Century," *Scientific American*, Special Issue, 1995.

Base Technology

Andersen Consulting, *Trends in Information Technology*, McGraw-Hill, 1993.

Lucky, R., *Silicon Dreams: Information Man and Machine*, St. Martin's Press, 1991.

Networks

Atkins, J., and M. Norris, *Total Area Networking: ATM, SMDS and Frame Relay Explained*, John Wiley, 1998.

Quarterman, J., and C.-M. Smoot, *The Internet Connection: Systems Connectivity and Configuration*, Addison Wesley, 1994.

Norris, M., and N. Winton, *Energize the Network: Distributed Systems Explained*, Addison-Wesley Longman, 1996.

Stallings, W., *Networking Standards: A Guide to OSI, ISDN, LAN and MAN Standards*, Addison Wesley, 1993.

Software

Coulouris, G., J. Dollimore, and T. Kindberg, *Distributed Systems: Concept and Design*, Addison Wesley, 1994.

Elbert, B., and B. Martyna, *Client Server Computing*, Artech House, 1994.

Meyer, B., *Object Oriented Software Construction*, Prentice-Hall, 1995.

Norris, M., *Survival in the Software Jungle*, Artech House, 1995.

Sommerville, *Software Engineering*, John Wiley & Sons, 1998.

Organization

Gray, M., N. Hodson and G. Gordon, *Teleworking Explained*, John Wiley & Sons, 1993.

Johansen, R., *Groupware*, Free Press, 1991.

Handy, C., *The Age of Unreason*, Arrow, 1995.

Whittaker, J., P. Ward, and P. Griffiths, *Strategic Systems Planning*, John Wiley & Sons, 1991.

Online References

It would be wrong to imply that all wisdom has been consigned to paper. Nowadays, there is a vast amount of relevant information online—far too much to catalog in any detail. The few references below are good starting points. In the absence of a meaningful reference, a few words of explanation is attached to each one.

http://pclt.cis.yale.edu/pclt/default.html
A very comprehensive and authoritative guide to networks. It starts by explaining basics—TCP/IP and the theory of networking—and then moves on to catalog the various pieces of hardware and software that help you to get the most from your connection.

A similar offering that is suited to the Macintosh is located at http://web.nexor.co.uk/public/mac/archive/welcome.html.

http://gnn.com/gnn/gnn.html
One of the network navigators. This page is easy to follow and gives ready access to a whole host of information sources.

For a "definitive" text on the Internet, the following documents can be retrieved from is.internic.net (e.g., FTP is.internic.net).

- RFC-1208—Glossary of Networking Terms;
- RFC-1207—FYI: Answers to commonly asked "experienced Internet user" questions.

The files are stored at an anonymous FTP site, so they can be accessed by replying "anonymous" at the login prompt.

About the Author

ABCDEFGHIJKLMNOPQRSTUVWXYZABCDEFGHIJKLMNOPQRSTUVWXYZABCDEFGHIJKLMNO

Mark Norris is an independent consultant with over 20 years experience in software development, computer networks, and telecommunications systems. He has managed dozens of projects to completion from the small to the multimillion dollar, multi-site and has worked for periods in Australia. He has published widely over the last ten years with a number of books on software engineering, computing, technology management, telecommunications, and network technologies. He lectures on network and computing issues, has contributed to references such as Encarta, is a visiting professor at the University of Ulster, and is a fellow of the IEE. Mark plays a mean game of squash but tends not to mix this with other forms of interfacing. Mark can be found at mnorris@iee.org.

Recent Titles in the Artech House Telecommunications Library

Vinton G. Cerf, Senior Series Editor

Access Networks: Technology and V5 Interfacing, Alex Gillespie

Advanced High-Frequency Radio Communications, Eric E. Johnson, Robert I. Desourdis, Jr., et al.

Advances in Telecommunications Networks, William S. Lee and Derrick C. Brown

Advances in Transport Network Technologies: Photonics Networks, ATM, and SDH, Ken-ichi Sato

Asynchronous Transfer Mode Networks: Performance Issues, Second Edition, Raif O. Onvural

ATM Switches, Edwin R. Coover

ATM Switching Systems, Thomas M. Chen and Stephen S. Liu

Broadband Network Analysis and Design, Daniel Minoli

Broadband Networking: ATM, SDH, and SONET, Mike Sexton and Andy Reid

Broadband Telecommunications Technology, Second Edition, Byeong Lee, Minho Kang, and Jonghee Lee

Client/Server Computing: Architecture, Applications, and Distributed Systems Management, Bruce Elbert and Bobby Martyna

Communication and Computing for Distributed Multimedia Systems, Guojun Lu

Communications Technology Guide for Business, Richard Downey, Seán Boland, and Phillip Walsh

Community Networks: Lessons from Blacksburg, Virginia, Andrew Cohill and Andrea Kavanaugh, editors

Computer Mediated Communications: Multimedia Applications, Rob Walters

Computer Telephony Integration, Second Edition, Rob Walters

Convolutional Coding: Fundamentals and Applications, Charles Lee

Desktop Encyclopedia of the Internet, Nathan J. Muller

Distributed Multimedia Through Broadband Communications Services, Daniel Minoli and Robert Keinath

Electronic Mail, Jacob Palme

Enterprise Networking: Fractional T1 to SONET, Frame Relay to BISDN, Daniel Minoli

FAX: Digital Facsimile Technology and Applications, Second Edition, Dennis Bodson, Kenneth McConnell, and Richard Schaphorst

Guide to ATM Systems and Technology, Mohammad A. Rahman

Guide to Telecommunications Transmission Systems, Anton A. Huurdeman

A Guide to the TCP/IP Protocol Suite, Floyd Wilder

Information Superhighways Revisited: The Economics of Multimedia, Bruce Egan

International Telecommunications Management, Bruce R. Elbert

Internet E-mail: Protocols, Standards, and Implementation, Lawrence Hughes

Internetworking LANs: Operation, Design, and Management, Robert Davidson and Nathan Muller

Introduction to Satellite Communication, Second Edition, Bruce R. Elbert

Introduction to Telecommunications Network Engineering, Tarmo Anttalainen

Introduction to Telephones and Telephone Systems, Third Edition, A. Michael Noll

LAN, ATM, and LAN Emulation Technologies, Daniel Minoli and Anthony Alles

The Law and Regulation of Telecommunications Carriers, Henk Brands and Evan T. Leo

Marketing Telecommunications Services: New Approaches for a Changing Environment, Karen G. Strouse

Mutlimedia Communications Networks: Technologies and Services, Mallikarjun Tatipamula and Bhumip Khashnabish, Editors

Networking Strategies for Information Technology, Bruce Elbert

Packet Switching Evolution from Narrowband to Broadband ISDN, M. Smouts

Packet Video: Modeling and Signal Processing, Naohisa Ohta

Performance Evaluation of Communication Networks, Gary N. Higginbottom

Practical Computer Network Security, Mike Hendry

Practical Multiservice LANs: ATM and RF Broadband, Ernest O. Tunmann

Principles of Secure Communication Systems, Second Edition, Don J. Torrieri

Principles of Signaling for Cell Relay and Frame Relay, Daniel Minoli and George Dobrowski

Pulse Code Modulation Systems Design, William N. Waggener

Signaling in ATM Networks, Raif O. Onvural and Rao Cherukuri

Smart Cards, José Manuel Otón and José Luis Zoreda

Smart Card Security and Applications, Mike Hendry

SNMP-Based ATM Network Management, Heng Pan

Successful Business Strategies Using Telecommunications Services, Martin F. Bartholomew

Super-High-Definition Images: Beyond HDTV, Naohisa Ohta

Telecommunications Deregulation, James Shaw

Telemetry Systems Design, Frank Carden

Teletraffic Technologies in ATM Networks, Hiroshi Saito

Understanding Modern Telecommunications and the Information Superhighway, John G. Nellist and Elliott M. Gilbert

Understanding Networking Technology: Concepts, Terms, and Trends, Second Edition, Mark Norris

Understanding Token Ring: Protocols and Standards, James T. Carlo, Robert D. Love, Michael S. Siegel, and Kenneth T. Wilson

Videoconferencing and Videotelephony: Technology and Standards, Second Edition, Richard Schaphorst

Visual Telephony, Edward A. Daly and Kathleen J. Hansell

World-Class Telecommunications Service Development, Ellen P. Ward

For further information on these and other Artech House titles, including previously considered out-of-print books now available through our In-Print-Forever® (IPF®) program, contact:

Artech House
685 Canton Street
Norwood, MA 02062
Phone: 781-769-9750
Fax: 781-769-6334
e-mail: artech@artechhouse.com

Artech House
46 Gillingham Street
London SW1V 1AH UK
Phone: +44 (0)171-596-8750
Fax: +44 (0)171-630-0166
e-mail: artech-uk@artechhouse.com

Find us on the World Wide Web at:
www.artechhouse.com